Project Adventure, Inc.

Ropes Course Safety Manual

An Instructor's Guide to Initiatives, and Low and High Elements

Steven E. Webster

A Project Adventure Publication in cooperation with

KENDALL/HUNT PUBLISHING COMPANY
2460 Kerper Boulevard P.O. Box 539 Dubuque, Iowa 52004-0539

WARNING

IMPROPER USE OF THE ADVENTURE ACTIVITIES DESCRIBED HEREIN MAY RESULT IN SERIOUS INJURY. THE ACTIVITIES SHOULD NOT BE ATTEMPTED WITHOUT THE SUPERVISION OF TRAINED AND PROPERLY QUALIFIED LEADERS.

NEITHER THE AUTHOR, PUBLISHER, SELLER OR ANY DISTRIBUTOR OF THIS PUBLICATION ASSUMES ANY LIABILITY FOR LOSS OR DAMAGE, DIRECT OR CONSEQUENTIAL TO THE READER OR OTHERS RESULTING FROM THE USE OF THE MATERIALS CONTAINED HEREIN, WHETHER SUCH LOSS OR DAMAGE RESULTS FROM ERRORS, OMISSIONS, AMBIGUITIES OR INACCURACIES IN THE MATERIALS CONTAINED HEREIN OR OTHERWISE. NO WARRANTIES, EXPRESS OR IMPLIED, AS TO MERCHANTABILITY OR AS TO FITNESS FOR ANY PARTICULAR USE OR PURPOSE ARE INTENDED TO ARISE OUT OF THE SALE OR DISTRIBUTION OF THIS PUBLICATION AND THIS PUBLICATION IS SOLD "AS IS" AND 'WITH ALL FAULTS'. THE LIABILITY OF THE AUTHOR, PUBLISHER, SELLER, OR ANY DISTRIBUTOR OF THIS PUBLICATION ON ACCOUNT OF ANY SUCH ERRORS, OMISSIONS OR AMBIGUITIES, SHALL IN ANY EVENT, BE LIMITED TO THE PURCHASE PRICE OF THIS PUBLICATION.

ISBN 0-8403-6207-2

Printed in the United States of America
10 9 8 7 6 5 4

Contents

Project Adventure, Inc.

Preface

Over the past 18 years, Project Adventure has accumulated a substantial amount of information regarding the safe operation of Adventure programs — including ropes courses and Initiatives. Two years ago we recognized the fact that the field of Adventure Education, which had grown significantly across the country, had no comprehensive body of material covering basic, safe operating procedures on ropes courses. This manual is the result of that recognition. Our purpose in writing it is to share our accumulated knowledge and experience and to promote safe, effective programs.

While the manual defines Project Adventure's minimum acceptable standards, and as such, the minimum standards for a PA accredited program, it should also be useful to other Adventure programs outside of our direct network.

Special thanks to all Project Adventure staff, trainers, and others whose input and constant striving to improve the quality of Adventure programs helped make this manual a reality.

Project Adventure, Inc.

Elements and Initiatives

Low Elements

Initiatives

High Elements

Introduction

Ropes Course — a definition

A challenge ropes course is a series of individual and group physical Challenges that require a combination of teamwork skills and individual commitment. Constructed of rope, cables, and wood, courses are constructed outdoors in trees (or using telephone poles) and indoors in gymnasiums.

Background

There were few ropes courses in existence when the first Project Adventure course was constructed in 1971 at Hamilton-Wenham Regional High School, in Hamilton, Massachusetts. Each Outward Bound School had a ropes course that was used (for one day at most) as part of the standard 26 day course. A few outdoor education centers, schools, and colleges had courses, which were usually constructed by Outward Bound alumni.

In 1974, the Project Adventure Program was awarded National Demonstrator Status by the Office of Education, and the program was given funds to disseminate the curriculum. The tenth grade physical education curriculum at Hamilton was and still is a year-long sequence of cooperative games and warm-ups, Initiative problems, and includes a low and high ropes course. *Cowstails and Cobras*, written and published by Project Adventure in 1977, and recently revised, outlined this curriculum and featured the use of the ropes course. The biweekly use of the course over much of the year was one of the most innovative aspects of the original PA program.

From 1974–1982, as part of the Hamilton-Wenham School District PA program, and since 1982 as part of Project Adventure, Inc., the PA staff has been helping others design, train for, and install Challenge ropes courses. To date there are, by most estimates, between two and three thousand courses in the United States, 1000 of which were constructed by PA staff. Most of the others have been constructed by organizations using their own staff and initiative. Individual entrepreneurs have increasingly entered the ropes course construction field.

The use of ropes courses has spread from the physical education classes and outdoor education centers of the early days, to a broad range of educational, recreational, and therapeutic organizations and institutions. Many camps now offer a ropes course station as a one to five day option for campers, and numerous Colleges offer ropes course programs as part of student leadership training, orientation programs, or as part of teacher education for the Adventure field. From the beginning, youth-at-risk programs and agencies have seen the potential of the high impact training offered by the ropes course. Increasingly, in the eighties, therapeutic agencies, substance abuse centers, and psychiatric hospitals have begun to use an Adventure program

known as Adventure Based Counseling. It would be safe to say that ropes courses have been used with virtually every type of population in-need in the country today.

Another recent trend is the use of Adventure programming in professional and corporate training. Team building needs, communication issues, creative problem solving, risk taking strategies, stress reduction techniques, and many other professional development themes have been addressed through Adventure Education approaches, which often use the ropes course as a training tool.

In the years since 1973, when PA staff first started teaching others how to replicate their model, the standards of construction, spotting, equipment, and belaying have undergone significant change. For example, in 1973 Hamilton students did the Zip Line on hand loops only and were belayed exclusively by a bowline around their waist. Both of these techniques have changed, as this manual indicates. In addition, the 10 and 15 year safety studies coordinated by PA, while showing a level of program operation safer than standard secondary school physical education classes, have affected the recommendations and standards agreed on for this manual.

Use of the Manual

This manual is intended to convey guidelines for the operation of a Challenge ropes course as used by Project Adventure staff and certified trainers. This manual will also serve as a basic guide for those programs seeking PA accreditation (see appendix A). PA will also use the manual as a basic operating guide for courses which we construct. Because of this, the manual has spaces designated for site-specific recommendations. For example: if there is a stump next to a Tension Traverse, the page covering the Tension Traverse has room to add additional site-specific instructions for this situation.

We also expect and encourage that this manual will be used as a guide for other programs seeking information about Project Adventure standards. Although the manual is intended to be generic, it is generic only for the large number of courses that use PA programming and whose staff members have received PA training.

Chapter 1

General Guidelines

Programs and Curricula — The Development of Goals

The establishment of goals on a ropes course enriches the interchange between the ropes course leaders and the individuals participating on the course. These goals should be clearly defined and, if possible, written.

The original Project Adventure (1971) student goals are:

1. **To increase the participant's sense of personal confidence.**

   The aim of many activities is to allow the students to view themselves as increasingly capable and competent. By attempting a graduated series of activities, which involve physical or emotional risk, and succeeding (or sometimes failing) in a supportive group atmosphere, a student may begin to develop true self-esteem.

   The curriculum has been planned to encourage students to try new and different activities — some of which may involve initial anxiety. It is our belief that as a person matures, he/she needs to learn to be familiar with the anxiety that precedes any new venture, cope with that uncertainty, and dare to enter fully into new situations.

2. **To increase mutual support within a group.**

   The curriculum is based on the assumption that anyone who conscientiously tries should be respected. Success and failure are less important than making an effort. In many cases the success or failure of a group depends on the effort of the members. A cooperative, supportive atmosphere tends to encourage participation. The use of teams, points, and timed competition, has been consciously minimized.

3. **To develop an increased level of agility and physical coordination.**

   A number of exercises entail the use of balance and smoothly flowing movement. Balance and coordinated movement form the basis for many physical activities ranging from dancing to track and football. A person who perceives himself as physically awkward, often sees himself as inadequate in other ways. Balance activities which can be successfully completed often give a feeling of accomplishment and personal worth to the doer.

4. **To develop an increased joy in one's physical self and in being with others.**

   One of the criteria which we have used in assessing various activities is that it must have a substantial element of fun in it. Instructors are not solemnly engaged in building confidence, social cohesion, and agility.

Just as people in the program may regularly be anxious and even fearful, so should they even more regularly experience joy, laughter, anticipation.

The program is designed to give the participants many opportunities to come to recognize that they are physical beings and that using one's body can be a joyous, satisfying, exhilarating, and unambiguous experience.

5. **To develop an increased familiarity and identification with the natural world.**

   Young people too often have little experience with sun, rain, dirt, snow, cold, Spring, Fall, and Winter. Because of the activities involved in the program, the students become increasingly comfortable with rolling in the dirt, with the smell of the grass, with the feel of rain or snow or wind, of cold or sunlight — in all their various moods. The weather always adds to the unpredictability of the chosen activity.

These goals were written for a tenth grade interdisciplinary public school program that involved physical education and academic classes. Modifications of the original school model have occurred in a variety of settings using a PA program approach, and have produced a large number of modified curricula. The Project Adventure publication *Cowstails and Cobras II*, a revision of the original *Cowstails and Cobras*, gives physical education and recreation curriculum examples, while another PA publication, *Islands of Healing*, gives examples of therapeutic curricula and approaches. It is important for all those using a course as a staff member to understand how, and towards what programmatic goals, the course is being used.

In general, whatever the curriculum approach, staff need to address the following questions regarding overall program design in their use of the course:

- Is there an adequate understanding by both instructional staff and program participants of program goals?
- Is there an adequate briefing–leading–debriefing strategy in place?
- Is there adequate provision for a range of population types and abilities that may use the curriculum and course?
- Are there strategies in place to insure transfer of the active-learning on the course to the other parts of the curriculum?
- Are the elements and activities appropriately sequenced to achieve program goals?

The program manager or department head in charge of the course, or curriculum that includes the ropes course, needs to review continually the above questions and manage the program so that these questions can be answered in the affirmative.

Management Issues

In most organizations there are various chains of command where people are answerable to others, and in the better ones there exists a flow of information among all parties. In effective organizations, individuals understand their own responsibilities. Because of the fact that a ropes course may be used by more than one class or institutional group, it is important that the roles of individual staff members and lines of authority are clear. There should always be one person whose responsibility is to see that roles are defined and understood and that management policies are followed. Program policies should address the following questions:

- What are the training and experience requirements for program leaders, and are these documented?
- How and when is the course to be inspected for physical integrity and safety by your staff? — by an external source?
- Will outside groups be allowed to use the course and if so, under what policies?
- Do you have an on-going periodic review of procedures for current staff?
- Are you using your manual in staff training to ensure that all staff members are familiar with current procedures and that those procedures are being followed?
- Do you keep written records of program use of the course and equipment?
- Do you have first aid equipment available during all programs and access for emergency vehicles in case of an accident?
- Is there a need for an emergency rescue plan? If so, is at least one staff member present at all times who can perform rescues?
- Do you have a liability release form for all participants?
- How does the program keep records of accidents and near misses?

It is suggested that programs have written policies which address these and other management questions. The answers which your program develops will vary depending on the size and scope of the program. While this manual is intended to help existing and beginning programs answer many of their questions, it is in no way intended to replace the necessary training or customized review of local policies and procedures that a PA accreditation visit can provide.

Training Issues

One of the principle missions of Project Adventure is to provide training to others who wish to implement a Project Adventure program at their own institution. The minimum required training for beginners is either the Adventure Programming or Adventure Based Counseling workshop. Both are five day training sessions which provide instruction in games, Initiatives, low and high ropes course skills, and curriculum design issues. At the end of the

workshop participants fill out a self-assessment form and receive feedback on their level of skill. Depending on their skill level and the design of their program, many participants are then ready to begin instructing in new or existing programs. Some individuals, however, may need more training, either through individual instruction at their site or further workshop training. Some also discover that the Project Adventure program, or at least parts of it, may be something which they should not lead.

Project Adventure's Accreditation Program (appendix A) provides an on-site service to program directors, instructors, and other interested parties to ensure that their equipment, program design, and instructor skills are within Project Adventure guidelines. The service provides professional feedback and identifies programs where more training may be necessary. PA programs vary greatly in design and complexity: from a simple ten week cooperative games and Initiative sequence, to a one day low ropes course experience, to a two year curriculum that involves hundreds of activities, including high and low ropes elements. Because of this variety, the determination that more training may be necessary can only be made on a case by case basis. The Accreditation Program is one method of accomplishing this.

Project Adventure's Advanced Skills Workshop is designed for those program leaders and instructors who wish to look at program management issues, current safety principles, and rescue procedures and plans. It is an experiential workshop that expands upon and reviews many of the introductory workshop skills. It is not intended for beginners. Project Adventure strongly recommends that programs have at least one person in a leadership position attend an Advanced Skills Workshop.

Like other skills, the skills of a Project Adventure instructor need to be practiced to be kept fresh and strong. Project Adventure recommends at least annual refresher training for all programs.

General Safety Principles

The following principles are intended to remind us all of the basics. Some might think that they are too obvious to include. It is our experience, however, that these principles are too often slighted or forgotten. Constant vigilance to adhere to these principles will help us all run safer and more effective programs for our participants.

Carefully sequence program activities.

This is important as much for safety reasons as it is for achieving the stated goals of your program. Safety has a great deal to do with having the right group on the appropriate activity at the right time. A single sequence is not appropriate for all groups and it may be necessary to change activities on a given day if a group is not ready for a particular activity.

Develop strong assessment skills.

The ability to choose the appropriate activities for a particular group requires strong assessment skills, which usually only develop with experience. For

example, there are some swinging elements which are very difficult to spot. In such cases the decision to use the activity is based upon the judgment of the instructor that the Challenge presented is appropriate and can be safely undertaken.

Don't economize on training.

Select instructional staff with care and don't economize on training or other safety related matters. Be prepared to stop running the program or cut out certain activities if staff are not properly trained.

Have your ropes course inspected.

A thorough annual safety inspection, as well as daily monitoring of the condition of elements and equipment, is highly recommended.

Begin slowly and carefully.

It is particularly important for new programs to begin slowly and carefully. It takes time and experience to develop the wide range of skills necessary to be safe and effective in leading groups. Start with small groups of cooperative students, and use co-leadership wherever possible.

Stay up-to-date with current safety procedures.

Schedule skill reviews and updates on an ongoing basis, and visit other programs whenever possible. See appendix C for a list of activities and procedures that are no longer used in Project Adventure workshops.

Set a good example.

The example we set is important for both safety and program reasons. If you ask students to wear a helmet, wear one yourself. Consider carefully the effect you may have if you climb unbelayed to set up the course. And if you ask your participants to make a good effort and to support each other, your personal example is certainly as important as what you say.

Be careful about doing something new and different.

Don't experiment with a new activity or untried procedure with students where safety issues are involved. New and different could be a new harness, a new self-locking carabiner, or a new activity. Or it might mean that a different population may encounter difficulties on a familiar activity; i.e., no one ever fell off the Burma Bridge before. Even a new solution to an Initiative problem might raise new spotting issues. Thoroughly evaluate all new procedures and activities before implementing them.

Weather and Course Conditions

"*If you don't like the weather, wait ten minutes*" (Old New England saying.)

Project Adventure programs are traditionally run outdoors and in all kinds of weather. Being outdoors can increase the Challenge and can be a positive group experience.

Because our New England roots make us particularly sensitive to changes that can occur in the weather, we've developed some guidelines. The most obvious threat to safety from adverse weather is from thunderstorms. Climbing around in trees with metal cables strung between them is not a place to be during an electrical storm. Extreme heat, cold, rain, snow, and ice can affect participants and instructors alike, and must be taken into consideration when planning and leading a group session.

- Listen to local weather reports and take them into account when planning for the day.
- Be aware of local weather tendencies (storms come from the east, storms come off the mountain, etc.), and pay attention to these possibilities.
- Err on the side of caution with thunderstorms, and leave an outdoor site when a storm threatens seriously.
- Staff should assure that all participants are dressed properly for weather conditions.
- If it is unusually hot, have water available, and take frequent breaks to assess participants for symptoms of dehydration.

Chapter 2

Introductory Activities

Warm-Up Exercises

Over the years, literally hundreds of warm-ups, games, and movement activities have been developed by Project Adventure, The New Games Foundation, and other similar organizations. Effective use of warm-up exercises helps set the tone for trying in a supportive, non-competitive atmosphere.

The actual sequence of warm-up activities varies because of differences in length and goals of individual programs. However, introductory activities should, where possible, be used to achieve the following goals:

- Development of trust: The appropriate sequencing of activities is vital to the development of group trust, when the group functions together and takes care of its members.
- Development of group cooperation and initiative: The ability to cooperate in a fun and engaging way is a theme of many of the most effective warm-ups, and in turn, reinforces the trust level and problem solving skills of the group.
- Development of a sense of body and awareness of movement : Warm-up activities — which run the gamut from, throwing, falling and catching, running, and rolling, to mirroring anothers' movements — are all potentially valuable experiences which reinforce the sense that participants can have fun, learn, and take care of each other, all at the same time. The sense of potential body movement learned in these program situations helps foster good spotting techniques.

Spotting

Spotting is a human safety net provided by other people for the person(s) doing an activity. It is the primary safety system for Initiative problems and low elements. It is also used in limited ways for certain high elements. Basic spotting techniques are taught to participants through introductory activities and then modified or added to as the demands of specific elements require.

Falling off things is to be expected. Proper, alert spotting helps prevent falls from causing injury. Regardless of the specific spotting technique being used (techniques do vary for different events and elements), the primary principle is to protect the participant's head and upper body through physical support.

The term *catching* is often used in both the teaching and doing of spotting. But while spotters usually need to be in position to catch a participant in the event of a fall, it is very difficult to literally *catch* a falling body, even from just a small height. Spotters and participants must understand that catching means to support and protect the upper body and head in case of a fall (sometimes referred to as *breaking the fall*). An understanding of this concept of catching, along with a grasp of the real meaning and function of spotting, can prevent breaking the initially fragile feelings of trust. Participants should understand that a spotter who breaks their fall, even though either or both participant and spotter end up on the ground, has indeed performed her/his role responsibly and in a trustworthy manner. Falls, along with minor scrapes and bruises, come with the territory.

Teaching spotting is one of the most important tasks in any ropes course experience. Careful instruction is required because potential spotters oftentimes do not recognize their importance until actually called upon to protect someone.

Here are some key aspects to remember in teaching spotting to a group:

1. Effective spotters follow the movements of the participant doing the activity — especially in the case of traversing elements; i.e., Tension Traverse, Swinging Log, etc. By paying close attention to the participant's movements, the spotter is forced to follow whatever movement the participant makes and positions the spotter to respond to a fall.
2. Spotters need to be able to move in and dampen any motion in a swinging activity; e.g., when swinging onto a Swinging Log, landing on the Seagull Perch, or Nitro Swing.
3. Develop a sequence for teaching the spotting of each element.
4. Practice spotting using activities designed to reinforce proper technique; i.e., Partner Trust Fall, Three Person Trust, Trust Circle, Two Line Trust Fall, and Gauntlet.
5. Be clear in explaining how good spotting enhances and develops trust among participants. Joking about *not catching* someone has no place on a ropes course.
6. Carefully distinguish the differences, among spotting, catching, and assisting.
7. Supervise spotters closely, reminding them of proper technique as needed.
8. Rotate spotters so everyone has a chance to spot and is used to spot other appropriately sized participants. (The big people should not end up doing all the spotting.)

9. Pay close attention to the number of spotters required to spot each element effectively. Size, strength, weight, fatigue, and group dynamic factors may also affect minimums. Do not hesitate to require more than the stated minimums in particular situations. But remember that too many spotters spotting for the same situation can lead to the problem of no one taking her/his job seriously enough because nobody feels their role is important.

Chapter 3

Low Element Guidelines

The group's first experience on a ropes course usually involves one of the low elements or an Initiative problem. A low element is an unbelayed activity that focuses the Challenge on one individual's achievement, while the other members of the group act as spotters. An Initiative utilizes the resources of the group to solve a problem together. The line between a low element and an Initiative is often a little fuzzy, and some elements can be used either way. The specific activity, and the manner in which you choose to use it, will depend largely on your particular program and the predetermined goals for the group. In the original PA physical education curriculum, low elements were done earlier in the curriculum sequence than Initiatives. But many programs today use only Initiatives. A careful review of low elements may help you pick up effective spotting strategies and assist you in your Initiative presentations.

Because of the number of activities available, ropes courses will vary greatly from one to another. To help instructors identify and understand the potentially different spotting techniques unique to each course, we have set aside sections under each activity for Comments or Special Considerations to be added by instructors, allowing each manual to be made site specific.

Each low course activity has been divided into three sections:

1. **Instructor's Role**
2. **Spotters Knowledge**
3. **Participants' Responsibility**

This approach provides an overview of the behavior and awareness that instructors, spotters, and participants should have to ensure a safe and effective program. It is not, however, the intention of this manual to require *one way* of presenting an activity or to dictate leadership style. There are many different styles and a diversity of presentation methods which can be effective.

Despite the variety of leadership styles, however, there remains the leadership responsibility to manage the activity in such a way that learning and safety are maximized.

All PA staff are asked to utilize the Full Value Contract with participants as a way of engaging themselves and the group in monitoring their own safety. In its simplest form the Full Value Contract is an agreement by all group members that they understand fully the safety and spotting rules as given by the leader, and that they will adhere to those rules. In addition, group members agree to speak up if they feel another member of the group is not following the guidelines. This simple agreement serves to start the process of getting the group to become more safety conscious and responsible for its own members. The Full Value Contract has many other program functions,

especially in therapeutic programming, but its safety value is basic. Depending on the group, the contract needs to be reinforced and revisited periodically.

To help in determining the technical skills necessary (degree of spotting difficulty), we have labeled each of the low activities and Initiatives using the international ski trail symbols as a means to determine the spotting skills necessary to safely process a group through the task.

Beginning — ●

Intermediate — ■

Advanced — ◆

These symbols also serve to remind the instructor to remain diligent and to keep in mind spotting considerations when designing or altering a program activity sequence.

The Low Course — Activities Available

Criss Crotch ●
Wild Woosey ●
Trust Fall ■
Single and Triangular
Tension Traverse ■
Fidget Ladder ■
Swinging Log ■ ***& Swing*** ◆
Bosun's Chairs ◆
Hickory Jump ◆
Swinging Tires ◆
Seagull Swing ◆

Criss Crotch ●

Two tautly strung and intersecting cables located close to the ground, upon which individuals traverse from one support tree to the other.

Group Size:

One or two persons can be on this activity at a time. The minimum number of spotters is two per participant. Other spotters may be needed depending on the type of Challenge presented.

Instructor's Role:

A. Inspect area for unsafe ground cover and litter.

B. Inspect trees for soundness, security of cable attachment points, and check for frayed cable ends.

C. Present the task, and review spotting requirements.

Spotters' Knowledge:

A. How the group will spot this activity.

B. The understanding for the potential of forward or backward falls and how to respond to either.

C. How to move laterally to break a fall.

Participant's Responsibility:

A. To ask for additional spotters if they feel it's necessary.

B. To inform spotters when they are going to try something different.

Variations:

Have two participants start the activity at opposite ends and pass one another in the middle or where the cables intersect.

Comments or Special Considerations:

Wild Woosey ●

Two tautly strung cables, close to the ground and starting at the same point, progress outwards to end points approximately 10–12 feet apart. Two participants (one per cable) walk the diverging cables, while maintaining physical contact, to a point where they can no longer continue or they reach the far support trees.

Group Size:

Two participants with a minimum of two spotters per participant and more added beneath the traversing pair as they progress outwards from the starting point.

Instructor's Role:

A. Check area for unsafe ground cover or litter.

B. Inspect trees for soundness, security of cable attachment points, and check for frayed cable ends.

C. Present the task, and review spotting requirements.

Spotters' Knowledge:

A. Understand that participants may fall in any direction when first mounting the cable.

B. *Do not* allow participants to interlock fingers while attempting the activity.

C. Spotters beneath the participants should always have hands clasped on top of their knees to prevent sudden backloading by falling participants.

D. Spotters beneath should only move as fast as the participants.

E. Add spotters as needed.

F. Understand that spotters located outside the cables are particularly important for the first ten–fifteen feet of the activity.

Participant's Responsibility:

A. Do not interlock fingers while attempting this activity.

B. Inform spotters when beginning the activity.

C. Communicate with partners.

D. Be aware of undue pressure on your partner's wrists and exhibit compassion by stepping down if the pressure causes pain.

Variations:

A. Establish a point that participants will go to and return from.

B. If pairs make it to end, attempt a return to the beginning.

C. Create a group problem: add together the distance traversed by each pair to create a group goal.

Comments or Special Considerations:

Trust Fall ■

A platform or stump ranging from 3 to 5 1/2 feet above the ground from which an individual falls backwards into the arms of spotters.

Group Size:

One person doing the falling (faller), with a minimum of eight catchers (spotters).

Instructor's Role:

A. Check area for unsafe ground cover; i.e., broken glass, dead limbs, stumps, etc.

B. Inspect stumps or platform for soundness.

C. Present the task, and review all spotting requirements.

D. Never be the first person to fall.

E. Be sure the group goes through a series of warm-up exercises leading into this activity.

F. Only do this activity when a group is ready.

G. Make sure all jewelry, watches, pencils, and pens are removed from all participants.

Spotters' Knowledge:

A. Clear understanding of what the activity entails from their standpoint as spotters and to buy into this experience.

B. Understanding of the strongest catching position:

- Knees are flexed.
- Arms are extended and bent at the elbows with palms facing up.
- Feet situated so that if a spotter is on the left side (facing the platform or stump), his/her right foot is extended in front of the left by 10 to 12 inches.
- Feet are at shoulder width.
- Arms alternate with the spotter opposite, and each spotter's fingertips extend to the opposite spotter's elbows.

C. Attention is focused on the faller at all times.

D. Upon catching a faller, spotters need to pay close attention to the faller until she/he is in an upright, standing position on the ground, and, if needed, the spotters should assist the faller to that position.

E. Spotters communicate with the fallers and let them know when they're ready.

Participants' Responsibilities:

A. Agree with the spotters on a series of commands by which fallers alert spotters of their intention to fall, and only to fall when spotters give the all clear by saying, "Fall."

B. Maintain a rigid position with head slightly back. Fallers draw their hands, fingers interlocked, into their chests in a firm grip, or use some alternate hand fixing system.

C. Line up at the direction of a spotter and fall without going into a pike position.

D. Upon falling and being caught, fallers allow themselves to be raised to an upright position, keeping their knees slightly flexed.

Variations:

Comments or Special Considerations:

Single Line and Triangular Tension Traverse ■

A single cable, or series of three cables, are strung tautly between trees at a height of no more than 2 1/2 feet above the ground and form a single line tension cable (one) or triangulated tension cables (three). An additional cable and rope (or cables) are suspended at the beginning of the single cable, or at the apex of the triangulated cables, to provide a tensioned rope to use while traversing the activity.

Group Size:

One participant with a minimum of two spotters on the single line, and two participants on the triangulated cables with two spotters for each participant as a minimum.

Instructor's Role:

A. Check area for unsafe ground cover; i.e., broken glass, dead limbs, stumps, etc.

B. Inspect trees for soundness, security of cable attachment points, and check for frayed cable ends.

C. Present the task, and review spotting requirements.

D. Make sure the suspended rope, or ropes, are long enough, especially if a transition step is to be made by participants at the corners.

E. Make sure the rope is free of ground litter and knots.

Spotter's Knowledge:

A. A clear understanding of how participants will move on this activity and an overall preparedness to spot a falling participant.

B. A recognition that when falls occur, they usually happen quickly and will invariably pull the participant back towards the starting point.

C. The most effective position for spotting is approximately a half step back towards the starting point from where the participant is then located on the cable.

D. A minimum of two spotters is necessary, but your assessment may require four spotters — two on either side of the cable.

E. If a fall occurs, spotters need to move in towards the participant and not back away.

F. Spotters need to help keep the tension rope free from snagging.

G. Spotting is necessary at the corners of the the Triangular Traverse and will involve more spotters.

H. Be aware when spotting on the back cable in the Triangular Traverse, that participants will be leaning back and that this position can allow uncontrolled fast movements laterally to the right or left.

I. Spot for a person coming back on the last cable. Be alert — participants regularly fall off this section.

Participants' Responsibility:

A. Ask for assistance and spotting when mounting or dismounting this activity.

B. If a fall occurs, simply stepping off of the cable can help prevent a participant from swinging towards the starting point.

C. Do not jump off of the cable in an attempt to regain balance.

Variations:

A. On a single cable out and back.

B. On the triangular version, pass around another participant on the back leg.

C. Do the activity over a clear swamp or water area and eliminate the spotters.

Comments or Special Considerations:

Fidget Ladder ■

A diamond shaped rope and wood ladder strung at an oblique angle between two points. Using hands and feet for balance, an individual attempts to maneuver, starting at the low end, about 2 1/2 feet above the ground, to the top of the ladder, usually 9 feet above the ground.

Group Size:

Only one participant on the ladder with a minimum of four spotters.

Instructor's Role:

A. Check area for unsafe ground cover; i.e., broken glass, dead limbs, stumps, etc.

B. Inspect ladder for loose lashings on rungs, and check rungs for cracks.

C. Attach and pull the ladder to proper tightness.

D. Present the task, and review spotting requirements.

Spotters' Knowledge:

On this particular activity a number of different spotting styles are used. We recommend four basic techniques:

1. The easiest but least used outdoors — placement of spotting mats under the ladder, with two spotters at the upper end to protect the participant's head and shoulders from hitting the support if a fall takes place at the top end of the ladder.
2. The use of all participants in the group spaced equally along the ladder with one arm extended over the ladder and the other extended below forming a cocoon.
3. The use of a small cargo net or old shrimp net which runs the entire length of the ladder and extending two to three feet beyond the widest point of the ladder. All members of the group act as spotters and grasp the edge of the net, holding it under the Fidget Ladder. If a fall takes place, spotters step back, catching the participant in the net.
4. If the ground under the Fidget Ladder is dug up or soft enough, then the only spotters needed are at the top end.

Depending on which technique is used, there are general guidelines which must be followed:

A. One or two spotters stabilize the ladder until the participant is ready to begin.

B. Two spotters are *always* at the top end of the ladder, positioned with their bodies next to the solid support.

C. Spotters on the high end (against the tree or pole) need to have their inside hand raised to spot the neck, and their outside hand low, to protect the lower back of participants.

Participant's Responsibility:

Whenever a fall occurs, hold on with both hands.

Variations:

Let participants experiencing little success use their knees for support.

Comments and Special Considerations:

Swinging Log ■ *& Swinging Log with Swing* ◆

A log or telephone pole suspended between two trees or poles. The log can measure from 15 to 30 feet in length and is suspended no more than 14 inches off the ground. If a swing accompanies this activity, it should be considered an *advanced activity.* The swing is designed to allow a participant to push off a stump, rock, or knoll, swing in an arc, and land on the log, using the rope in an attempt to maintain balance on the log for 10 seconds.

Group Size:

One or two participants on the log at a time depending on what type of Challenge is being presented. Types of Challenge include:

A. One individual mounts and walks the log, attempting to maintain his/her balance.

B. From about 3 feet away from the log, attempt to step onto the log and maintain balance for 5 seconds.

C. Two participants start at opposite ends of the log and attempt to walk towards one another, pass around each other, and continue on.

D. Two individuals start at opposite ends of the log and try to unbalance one another through various log rolling techniques.

E. To swing from a stump, land on the log, gain balance, and hold onto the swing rope while waving a hat in the other hand. This usually is attempted after mastering the swing onto the log.

Instructor's Role:

A. Check area for unsafe ground cover; i.e., broken glass, dead limbs, stumps, etc.

B. Inspect trees and log for soundness, security cable attachment points, and check for frayed cable ends.

C. Present the task, and review spotting requirements.

D. Give a complete demonstration of how the log can move and the various arcs through which it can swing.

E. Have the group lift the log to feel its mass and weight, and potential for inertial movement.

Spotters' Knowledge:

A. Understand their spotting position at all times in relation to the log, know how the log will swing if a participant falls, and how to protect both themselves and the participant when a fall occurs.

B. Four spotters need to be in position to dampen the motion of the log after a participant falls or steps off the log.

C. How to move in to protect a backwards fall off the log.

D. Be able to move with the motion of the participant as well as the log.

E. Spotters never position themselves where they can be hit by the swinging log.

F. Additional knowledge for swinging activity:

- Move in and stop a swinging participant if they see that participant coming in sideways or backwards.
- To have special spotters assigned to prevent participant spotters from moving into support cables when dampening the motion of an out of control swinger.

Participants' Responsibility:

A. Never jump off the log, propelling it in the opposite direction of the jump and into the spotters.

B. Indicate to the spotters their intention; to swing onto the log or to step up onto it.

C. Do not attempt running on the log.

Variations:

Walk the log backwards.

Comments or Special Considerations:

Bosun's Chairs ◆

A series of steps suspended at varying heights from a cable, which an individual crosses by swinging from step to step.

Group Size:

It's a good idea for one person to do this activity with a minimum of two spotters moving with the participant.

Instructor's Role:

A. Check area for unsafe ground cover; i.e., broken glass, dead limbs, stumps, etc.

B. Inspect trees for soundness, and security of cable attachment points.

C. Present the task, and review spotting requirements.

D. Check each step for rot and sharp splinters, and make sure there is no severe cracking (which may cause a step to split).

E. Make sure knots tied under steps are secure.

F. Check ropes for wear — pay special attention to any rope made of polypropylene, whose average outdoor life is two years because of UV degradation.

Spotters' Knowledge:

A. A clear understanding of how to move with the participants as they swing from step to step.

B. Understand that their role is to be there in case of an unexpected slip of the foot or loss of strength in the arms and upper body of the participant.

C. Always spot with arms raised — an effective spotter always keeps his or her arms raised.

D. Not only be aware of participants, but be aware of swinging chairs.

E. An understanding of how to break a fall by supporting the participant's upper torso (chest, arms, head), and to recognize that if spotters grab the participant's feet or legs, the participant will tumble over head-first.

Participants' Responsibility:

A. To recognize the fatigue factor of the upper body (arms, shoulders, and hands), that strength can be sapped quite easily, and to inform spotters if this occurs.

B. To feel comfortable in asking for additional spotters.

Variations:

Comments or Special Considerations:

Hickory Jump ◆

A trapeze is affixed to two cables hanging about 8 feet above the ground. An individual starts on a series of graduated stumps (height and distance) directly back from and facing the trapeze, and dives forward in an attempt to catch the trapeze. Attempts continue from progressively distant stumps until the participant jumps and fails to catch the trapeze.

Group Size:

One jumper with a minimum of eight spotters.

Instructor's Role:

A. Check the area for unsafe ground cover; i.e., broken glass, dead limbs, stumps, etc.

B. Inspect support trees and stumps for soundness, security of cable attachment points, and check for frayed cable ends.

C. Present the task, and review spotting requirements.

D. To make sure all jewelry, pencils, large belt buckles, knives, etc., are removed from all jumpers.

Spotters' Knowledge:

A. A clear understanding of how to catch a diving person versus a falling one, as in the Trust Fall.

B. Understanding that the strongest catching position is slightly altered from the Trust Fall:

- Knees flexed.
- Arms extended and bent at the elbows with palms up.
- Furthest shoulder from the platform or stump leans in towards where the faller will land, thus extending this arm to allow faller to land in the crook of forearm and biceps.
- Feet situated so that if spotters are on the left side (facing platform or stump), their right foot is extended in front of the left by 10–12 inches. If this is done with the extension of the right shoulder, then the plane of the body will form a 45° angle to the line of the jumper.
- Feet are at shoulder width.
- Arms alternate with the spotter opposite.

C. Attention focused on the jumper at all times and always catch the jumper even if he/she catches the trapeze.

D. NEVER LET A JUMPER SWING THRU after making a catch; this is crucial for safety.

E. Upon catching a jumper, always assist him/her into an upright standing position.

F. Spotters communicate with the jumper to let him/her know when they are ready.

G. Spotters who are positioned the furthest distance from the trapeze (towards stumps), protect shins, ankles, and feet of tall persons and long commitment jumpers from hitting the stumps.

Participants' Understanding:

A. Agree with the spotters on a command which alerts spotters to your intention to jump. Also agree not to jump until the spotters give an all clear by saying, "Jump away," or another agreed upon signal.

B. Refrain from kicking or pumping legs and feet while in the air.

C. Ask for spotters while mounting or walking up the stumps.

Variations:

Comments or Special Considerations:

Swinging Tires ◆

A series of tires suspended from an overhead cable. Participants must swing from tire to tire to cross a designated area. This activity often begins with a swing rope to the first tire and ends with a swing from the last tire.

Group Size:

One or two participants with a minimum of two spotters if requested by the participant or judged necessary by the instructor.

Instructor's Role:

A. Check area for unsafe ground cover; i.e., broken glass, dead limbs, stumps, etc.

B. Inspect trees for soundness, security of cable attachment points, check for frayed cable ends near tires.

C. Present the task, and review spotting requirements.

D. Check tires and make sure they have drain holes for water and, if not, drain them.

E. Provide gloves if the tires are supported by cables to the suspension cable.

F. Alert participants that they should not swing in the tires with their heads hanging down below their buttocks.

Spotters' Knowledge:

A. A clear understanding of how to move with the participants.

B. An understanding that their role is to be there in case of an unexpected slip of a participant's foot or loss of strength in the arms and upper body of a participant.

C. Not only be aware of the participant, but be aware of swinging tires which the participant has left.

D. Competence in knowing how to break a fall by supporting the participant's body.

Participants' Responsibility:

A. Recognize the fatigue factor of the upper body (arms, shoulders, and hands), that strength can be sapped quite easily, and to inform spotters if this occurs.

B. To feel comfortable in asking for additional spotters.

Variations:

A. If this activity is used as an Initiative task, and a heavy object is to be brought across as a "bomb", make sure it weighs no more than 25 lbs. It should be less if the population is younger.

B. Participants who have completed the task or not yet begun can assist in spotting.

Comments or Special Considerations:

Seagull Swing ◆

A long pendulum swing where a participant takes off from a stump, log, or perch of some sort, and swings in an arc, eventually landing on an elevated perch two to three feet off the ground.

The difficulty of this element arises from the fact that as participants swing, they must rotate their bodies 180 degrees before landing.

Group Size:

One participant swinging at a time with a minimum of six spotters — two positioned at the bottom of the swing, two in front, and two in back of the landing perch.

Instructor's Role:

A. Check area for unsafe ground cover; i.e., broken glass, dead limbs, stumps, etc.

B. Inspect trees for soundness, security of cable attachment points, and check the stability of the platform.

C. Present the task, and review spotting requirements.

D. Clearly assess the group's ability to do a swinging activity, and be aware of each participant's upper body strength.

E. Not to attempt this activity if the group is not capable of advanced spotting.

F. Demonstrate this activity.

Spotters' Knowledge:

A. The spotters positioned at the apex of the swing must know how to move in and support the swinger's upper body if the swinger does not have the arm or upper body strength to hold his/her own weight.

B. Spotters located at the landing perch react together to prevent the swinger from hitting the perch with his/her shins or ankles.

C. After the participant has landed on the perch, spotters continue spotting until full balance is achieved by the participant and he/she is returned safely to the ground.

Participant's Responsibilities:

A. Approach the landing with knees flexed; do not land with locked knees.

B. Communicate clearly his/her intentions to the spotters throughout the activity.

C. Try swinging without a full run if there is any doubt of the participant's ability.

D. Understand that spotters will err on the side of caution.
E. Ask for spotting (loudly) if there is any doubt about making the swing.
F. Do not raise knees and kick out when trying to land on the perch.
G. Never let go of the rope until securely perched on the platform.

Variations:

Comments or Special Considerations:

Chapter 4

Initiative Guidelines

Almost all ropes course experiences incorporate a series of Initiatives. Over the years, these group problem solving activities have been called many different names — A.S.E. (Action Socialization Experiences), Teams Course Challenges, Group Problem Solving Activities, and Initiatives.

A problem is presented to a group (usually 8–16 members) whose task it is to utilize all of the group members in completing, or attempting completion, of a specified task.

There are many variables which determine how Initiative problems fit into an Adventure Programming Experience, and these need to be consistent with your pre-established goals and program philosophy. But it's important that all programs include the three stages of an Initiative Problem:

1. Briefing, or verbal presentation of the problem.
2. The physical attempt by the group of the problem.
3. Debriefing how they functioned in the Initiative attempt.

The instructor serves as a facilitator who presents the parameters of the task to the group, serves as a clarifier of questions, acts as a safety valve to monitor and make sure the group attempts the task safely, and as an encourager and supporter of the group's attempts. The instructor should not serve as an answer-man, telling the group how to do the task. Finally, the instructor should assist the group in assessing how well they functioned as a group and what behaviors the group, and individuals within the group, exhibited.

Guidelines for the Instructor:

1. Present the problem clearly, and be clear about what additional props or equipment are allowed.
2. Support the group's efforts without telling them how to accomplish the task.
3. Be patient.
4. Don't disengage from what is happening — pay attention, observe behaviors, and be positive.
6. Remember that a positive learning experience does not always mean successful completion of a task — no matter how much the instructor and/or the participants may want that to occur.

Initiative Activities Available:

Maze ●	***Prouty's Landing and Disc Jockeys*** ■
Spiders Web ●	***All Aboard*** ■
T.P. Shuffle ●	***Vertical Pole & Tire*** ◆
Trolley ●	***Swinging Tires with Bomb*** ◆
Mohawk Traverse ■	***Hanging Teeter Totter*** ◆
Rebirth ■	***Beam*** ◆
Nitro Crossing ■	***Wall*** ◆

The list of available Initiative activities comes from those which represent physical activities associated with constructed elements on a ropes course. There are many additional Initiatives which are either no-prop or limited-prop activities. For additional Initiatives see *Cowstails and Cobras II*, *Silver Bullets*, or "BOT's", by Karl Rohnke.

Each Initiative write-up includes a section at the end for Comments or Special Considerations. This has been included to allow programs to make any additions which it deems necessary to make the manual site specific.

After a brief physical description of the Initiative problem, four sub-categories have been included:

1. The Task is a brief description of the activity, and while this is one example, it is certainly not the only one. Variations to the activity have been included.
2. The second section, Instructor's Role, assumes that each instructor has the same responsibilities as earlier described in the low elements chapter.
3. Spotters Knowledge has not been defined separately in this sub-section because it is felt that the evolutionary process of a group preparing to attempt a task should include this knowledge. Clearly, the hope is that the sequential/developmental approach of engaging activities on a ropes course allows for this evolution. But it is recognized that many programs that offer rope course experiences may only have groups on their course for a short period of time. In these cases instructors may have to be more directed with a group to ensure that spotters knowledge becomes part of the group process in determining task attempts.
4. The Participants' Responsibility section is included because once again every individual in a group is not only responsible to the group but also to themselves.

Maze ●

Generally constructed in a wooded area, the Maze consists of several 1/4" to 3/8" diameter ropes strung at waist height in and around a series of trees to form a single entry/exit maze. Dead ends may be created by placing pressure-treated landscaping timbers in the ground at strategic positions where ropes can then be strung to available trees. The entry/exit should be just small enough for participants to squeeze through.

Task:

The instructor has the group blindfold itself — far enough from the maze so that they cannot form a visual map of it. Groups of participants are walked to the maze and placed in it by ducking under the ropes. Once everyone is in the maze, they must find their way out. At no time can they take both hands off the ropes.

Instructor's Role:

A. Pre-check the area to make sure that there are no branches broken off at eye level, "pungi sticks" protruding up from the ground, broken glass laying around, or any obstacles which could injure a participant.

B. Clearly present the problem, and answer questions before the group begins the task.

C. Assuage fears by indicating that the blindfolds can be removed at any time.

Participants' Responsibility:

To always have one hand on a rope and not to duck under any ropes.

Variations:

A. Have the group do the task non-verbally.

B. Present the task so that if individuals decide to do the activity on their own, they can do so.

C. If the task is done individually or in small groups, and if an individual or small group finds their way out, they can then return to help others but they can't take off the blindfold. The instructor needs to tell participants when they exit.

Comments or Special Considerations:

Spider's Web ●

A prefabricated web strung between two trees 10–14 feet apart made up of 14 to 17 open web sections. The web may be constructed from of a variety of small ropes, twine, and bungee cord. Many have been constructed to be removable. The top of the web should be no more than 7 feet in height.

Task:

To pass each member of the group through a separate web opening without letting any part of the body touch any part of the web. Once a member uses an opening, that section is closed to further passage. If a participant on the opposite side of the web touches, that person has to return to the beginning side, and the section which they went through remains closed. Participants cannot be passed over or under the web.

Instructor's Role:

A. Check the area for unsafe ground cover; i.e., broken glass, dead limbs, and pungi sticks.

B. Present the problem clearly, review spotting requirements, and answer questions before the group begins the task.

Participants' Responsibility:

A. To be an active member of the group.

B. To give support to fellow group members both physically (spotting and passing) and emotionally (supporting and contributing ideas).

C. To never drop or let go of a participant because someone touched the web.

Variations:

A. If *anyone* touches the web, the whole group returns.

B. If there are more people than spaces, one or two spaces can be chosen for two group members to pass through, but once chosen, these spaces cannot be changed.

Comments or Special Considerations:

T.P. Shuffle ●

A telephone pole or log, 20–30 feet long, either laying on the ground or raised about 1 1/2 feet off the ground.

Task:

Half the group starts at one end of the pole, and the other half of the group starts at the opposite end. Facing towards the center of the pole, each group must move to the opposite end of the pole without anyone touching the ground.

Instructor's Role:

A. Check the area for unsafe ground cover.

B. Present the problem clearly, and answer questions before the group begins the task.

Participants' Responsibilities:

A. Agree that if a participant steps or falls off of the log, that he/she does not pull off the whole group.

B. The group agrees to communicate among themselves when moving.

Variations:

A. Add a time limit and assign penalty seconds for anyone who falls off.

B. Do the problem where each group starts backwards and ends up backwards.

C. Do the activity non-verbally.

Comments or Special Considerations:

Trolley ●

Two relatively knot-free, non-warped 4"x4"s, no more than 12 feet in length. An equal number of 4 foot ropes are inserted through the boards at 12 inch intervals.

Task:

To have the group walk a prescribed course with their right feet placed on one of the 4"x4"s, their left feet on the other, and holding onto the ropes. If any member falls off, the group returns to the starting point to begin again.

Instructor's Role:

A. Check the area for obstacles: crossing uneven ground strains or can break the 4"x4"s.

B. Present the problem, and answer questions before the group begins the task.

Participants' Responsibility:

Agree to communicate with one another, have fun, and be supportive.

Variations:

A. Depending on age level, two-person and four-person trolleys can be very effective.

B. To have the group go up and over objects, and down slopes (gradual).

C. To have the group move backwards rather than forwards.

D. If an individual falls off, have that person return to his/her position but turn 180° to the rest of the group.

Comments or Special Considerations:

Mohawk Traverse ■

Usually five separate sections of tautly strung cable (it can be more). The cables are strung between trees or poles no more than one to two feet above the ground. Distances between trees or poles vary from 8 or 9 feet, to 20 feet plus. Each section of cable can have a series of different Challenges to it; i.e., Multivine, Tension Traverse, modified Wild Woosey, etc.

Task:

For the group as a whole to traverse the entire length of the series of cables without falling or stepping off. If someone falls or steps off of the cable, that person, or persons, returns to the starting point or end of the line.

Instructor's Role:

A. Check the area for unsafe ground cover.

B. Clearly present the problem, review spotting requirements, and answer questions before the group begins the task.

C. Use group members where spotting is necessary.

Participants' Responsibility:

A. Agree that if a fall is imminent, to step off the cable and not pull off other participants.

B. Agree to communicate among all group members.

C. Agree to refrain from individual attempts without communicating with the group.

Variations:

A. If a member of the group falls off, have that person return to wherever the end of the group is located, or to the beginning

B. Have the whole group return to the beginning of the problem if any one member falls off.

C. Set a predetermined number of falls before the activity begins and allow fallers to get back on to the same spot from where they fell. If the group exceeds that predetermined number of falls, use any of the variations above.

Comments or Special Considerations:

Rebirth ■

A truck tire is hung between two trees at a height of 4 1/2 to 5 1/2 feet above the ground. A one inch drain hole is drilled into the bottom of the tire to allow water drainage.

Task:

For a group to pass each participant through the tire. Once through the tire, a participant is not allowed to return to the start and help, except as a spotter.

Instructor's Role:

A. Check the area for unsafe ground cover.

B. Clearly present the problem, review spotting requirements, and answer questions before the group begins the task.

C. Remove twigs, leaves, and other debris from inside the tire.

Participants' Responsibility:

A. Agree to spot each individual passing through the tire until they are placed on the ground.

B. Agree to be in the correct spotting position to assist the first and last person through the tire. (See Trust Fall to review good spotting technique.)

Variations:

A. To go through the tire with only one participant touching the tire.

B. For an easier variation, remove the rule against assisting the last several members through the tire.

Comments or Special Considerations:

Nitro Crossing ■

A spliced loop swing rope is suspended from a limb or cable. "Trip wires" are positioned at the beginning and end of the problem, usually about a foot off the ground.

Task:

The group has to obtain the rope using any resource found within the group and cross the area bordered by the trip wires without touching the ground. At some point during the activity the group must carry a #10 can (or similar container) filled 7/8 full with water (nitro), without spilling a drop. Once across, participants can only spot — they cannot return to give any assistance. If a trip wire is knocked down, the whole group must return. If any water spills, the whole group must also return, but if a group member touches the ground while swinging, only that person must return.

Instructor's Role:

A. Check the area for unsafe ground cover.

B. Present the problem, review spotting requirements, and answer questions before the group begins the task.

C. Check the rope to make sure it is sound.

D. Spot the first couple of people swinging across the bordered area or until enough group members have crossed and can provide spotting.

E. Do not allow diving or jumping for the rope.

Participants' Responsibilities:

A. Agree not to use excessive force to swing members across the no-touch area.

B. Agree to encourage and support each person's swinging effort. Avoid coercing individuals into trying this activity.

Variations:

If any group member touches, everyone returns to the start.

Comments or Special Considerations:

Prouty's Landing and Disc Jockeys ■

Task:

This activity is essentially the same as the Nitro Crossing except that instead of landing in the safe area, participants land on a 3'x3' platform (Prouty's Landing) or a series of wooden discs or hula hoops (Disc Jockeys). The instructor's role is the same as in the Nitro Crossing.

Participants' Responsibility

A. Agree to position themselves on the board or discs so that the safety of all group members is considered, and to avoid piling participants in layers on the platform.

B. Agree to ask for additional spotting if balance is precarious on the landing area.

C) Recognize that landing on a disc may cause it to slide and to be prepared for that.

D. Group members agree to communicate their intentions when swinging towards the platform or discs, and to plan together for the effect that a swinging body may have on a crowded area.

E. Agree not to allow participants on each other's shoulders throughout this activity.

Variations:

Comments or Special Considerations:

All Aboard ■

A two-foot by two-foot platform, which can be either permanently installed in the ground, or built on 4"x4"s to provide portability.

Task:

To see how many group members can balance on the platform, without piling themselves on top of one another, for a period of five seconds. Participants cannot touch the ground with any part of their body.

Instructor's Role:

A. Check the area for unsafe ground cover.

B. Clearly present the problem, review spotting requirements, and answer questions before the group begins the task.

C. Insist on additional spotters if a group has participants on the shoulders of other members.

Participants' Responsibilities:

A. To be aware of the strength and body size of group members, and agree not to have members lifting, supporting, or being supported in a manner in which they are not comfortable.

B. Agree not to attempt a solution that does not provide the group with adequate spotting.

C. Agree to give assistance to any and all group members who might need it.

Variations:

Have the group discuss and develop a strategy and then try it non-verbally.

Comments or Special Considerations:

Vertical Pole & Tire ◆

A large diameter car tire (usually 15"–16 ") is placed over a pole or topped tree, which can vary in height from 11–13 feet. At the top of the pole a peg or cut-off limb remains that prevents the tire from being thrown off of the pole.

Task:

Utilizing all group members, remove the tire from the pole, place it on the ground, and then place it back on the pole.

Instructor's Role:

A. Check the area for unsafe ground cover.

B. Clearly present the problem, review spotting requirements, and answer questions before the group begins the task.

C. Make sure dirt, leaves, twigs, and water, are out of the tire's cavity before the activity begins.

D. Check soundness of pole.

E. Review with the group how to support another's weight in such a manner as to reduce strain on participants' backs.

Participants' Responsibilities:

A. Agree to carefully monitor, spot, and support all group members.

B. Evaluate footwear and remove if it will damage participant's clothing and/or shoulders.

Variations:

Use a tire cut in half (around the middle) for younger groups.

Comments or Special Considerations:

Swinging Tires ◆

A series of tires suspended from an overhead cable. Participants must swing from tire to tire to cross a designated area. This activity often begins with a swing rope to the first tire and ends with a swing rope from the last tire.

Task:

For the group to gain access to the opposite shore while crossing a hazardous waste dump. The group must transport a "bomb" and all group members to the opposite shore.

Instructor's Role:

A. Check the area for unsafe ground cover.

B. Inspect trees for soundness, security of cable attachment points, and check for frayed cable ends.

C. Present the task, and review spotting requirements.

D. Check tires and make sure they have drain holes for water and, if not, empty them.

E. Provide gloves if tires are connected by cables to the overhead cable.

F. Alert participants not to swing in the tires with their heads hanging below the tires.

G. Be aware of participants fatigue level, and spot if evidence of fatigue sets in.

Participants' Responsibilities:

A. Agree to assist and spot participants as they move onto the first swing and off of the last one.

B. Agree to inform the group of weariness and fatigue and ask for spotting.

C. Agree to inform participants on the tires, and other group members, of their intentions when moving across the activity.

Variations:

Use a weighted object (no more than 25 lbs.) as the bomb, and agree that it will be passed, not thrown.

Comments or Special Considerations:

Hanging Teeter Totter ◆

A 6"–9" diameter, 8' long pole or peeled log is suspended from a very large overhead branch or cable via a section of 1" diameter rope. A small rope forms a circle on the ground with a diameter slightly greater than the length of the log.

Task:

To balance eight people on the log with no part of any group member's body touching the ground within the rope circle. If the log touches the ground, the group starts over.

Instructor's Role:

A. Check the area for unsafe ground cover.

B. Inspect trees for soundness, security of cable attachment points, and check for frayed cable ends.

C. Check the rope for any fraying or deterioration.

D. Check the log for rot, especially if it has been disconnected from the rope and left on the ground.

E. Present the task, and review spotting requirements.

F. Demonstrate how the log can rotate and swing out of the area prescribed by the circle. Discuss the implications of this on safety and spotting.

Participants' Responsibilities:

A. To be constantly aware of what the log can do when an individual(s) mounts or dismounts it.

B. To assign people specifically to spot the log throughout the entire task.

Variations:

Comments or Special Considerations:

Beam ◆

A debarked log or smooth telephone pole, 10"–14" in diameter, is suspended between two poles or tires, 8–9 feet off the ground.

Task:

For the group to get from one side of the beam to the other without using the side support trees or poles. Once over the beam, you can only give assistance on the far side. Spotting is allowed and encouraged at all times.

Instructor's Role:

A. Check the area for unsafe ground cover.

B. Inspect trees or poles for soundness, security of cable attachment points, and check for frayed cable ends.

C. Check the beam for soundness.

D. Check all cable connection points — this includes cable clips, serving sleeves, and eye screws supporting the cable.

E. If the beam is lashed to the trees or poles, check the lashings for vandalism or deterioration.

F. Review proper spotting of neck and upper torso area.

Participants' Responsibilities:

A. Agree to support and spot individuals from the beginning of their effort to mount the beam, pass over the beam, and are safely on the ground.

B. Agree to have no more than two participants at a time on the beam helping others.

C. Agree to spot at all times the two members of the group on the beam.

D. Agree never to throw someone up and over the beam in a zealous effort to complete the activity quickly.

E. Agree that no one jumps down from the beam .

F. Participants agree not to kick out with their feet in an attempt to get themselves up and over, putting spotters at the risk of suffering from "foot in mouth disease."

Variations:

With adequate spotting, allow an individual to try and get over the beam alone — a harder task than it might seem.

Comments or Special Considerations:

The Wall ◆

A smooth surfaced wall, usually ranging from 9 to 12 feet in height, which may be nailed to 4, 4"x4" cross supports, or have a platform built out from the second 4'x4' from the top. A means of descent must be provided on the back side.

Task:

Using all its members, the group must get everyone up and over the wall starting on the smooth surfaced side. The group may have three persons on the top of the wall, assisting a fourth person up and over. The sides of the wall and support trees or poles cannot be used. Articles of clothing are generally not used but see Variations.

Instructor's Role:

A. Check area for unsafe ground cover.

B. Inspect trees and support braces for soundness. Check wall surface for deterioration.

C. Affix the descending rope behind the wall, or if it's a permanent attachment, check the rope for soundness.

D. Make sure that the top and face of the wall are smooth and free of splinters.

E. Make sure that no nails are protruding from the wall.

F. Check Wall related structures for soundness — platforms, braces, railings, support ropes, ladders.

G. Review spotting procedures, and remind the group of the importance of group spotting due to the height of the obstacle.

H. Secure a commitment from the group to pay particular attention to spotting the back side of the wall.

I. Stress the importance of spotting an individual throughout the entire task.

J. Stress proper lifting and support, especially when participants are standing on other participant's shoulders or are being lifted up to that position.

K. Gain a commitment from the group to pay particular attention to spotting the last two members, especially if they are to try a running and jumping attempt.

L. Do not allow the group to use belts, shoelaces, or other articles of clothing that might not support the heavier members of the group.

Participant's Responsibilities:

A. Agree to support everyone's effort and to spot each participant from the beginning of starting the wall, while on the top, and all the way down the back side.

B. Agree to have only three people on top of the wall and one in transition.

C. Agree to have an appropriate number of spotters on the front and back of the wall at all times with their focus being on spotting only.

D. Agree not to hang the next-to-last individual by the legs in order to reach the last member of the group.

Variations:

A. If a group has difficulty with height or arm strength among the members, allow the rope to be used for the first and last two participants.

B. Immobilize an appendage to simulate a broken arm, leg, etc.

C. Attempt the task non-verbally.

Comments or Special Considerations:

Chapter 5

High Element Guidelines

Belaying

The high elements on a ropes course are those which require a method of safety known as a belay system. The belay system uses rope, carabiners, and other specialized hardware to provide protection for anyone climbing higher than can be safely spotted from the ground.

There are two different belay systems employed on ropes courses; Dynamic Belays, and Self-Belays, sometimes referred to as a static belay.

Dynamic belays involve the use of a belayer who controls the safety rope to the participant. In this arrangement the belayer remains on the ground and when a participant is ready to be lowered to the ground, or falls off the element, the belayer is in a position to bring the person to the ground in a controlled, gradual descent. Under the heading of dynamic belays, Project Adventure regularly teaches two separate methods; the body belay, which uses the belayer's body to gain friction, and the Figure 8 and Sticht Plate, both of which are mechanical devices that provide the friction to catch a fall or control the descent of a participant.

A self-belay uses a 4–5 foot rope to clip-in a participant to the belay point from their harness or swiss seat. It is, for the most part, put into service after an individual has been dynamically belayed up to a high element. Ropes courses, with high elements connected in a continuous pattern, often utilize a self-belay. Project Adventure does not instruct in this technique or construct static belay courses except upon special request. The original PA program utilized a dynamic belay in order to have a greater degree of control over the participant for safety reasons and to emphasize the trust aspect of the belay process. We have continued this tradition.

Basic belaying technique is taught and practiced in a systematic progression beginning with ground practice (on instructors and trainees). A backup system is used until competence and confidence are acquired.

> ***The descriptions in this chapter are not to be used in any way as a substitute for training, but as a means of review after training has occurred.***

General Guidelines for Belaying

Whether using a body belay or a mechanical belay device, these guidelines should be followed:

1. ALWAYS keep the braking hand in contact with the rope.
2. Stand as close to an element as space and conditions permit.

3. Proper tightness of the belay rope should be maintained. Keep excess slack out of the rope.
4. Keep even with any participant who is traversing an element. Move at the participant's pace and be ready to catch a fall at any time.
5. Make sure that the participant does not climb (up or down) faster than you can take up (or pay out) rope.
6. Keep full attention focused on the climber at all times.
7. Use proper climbing signals or some other agreed upon method to maintain good, clear communication with the climber at all times.
8. Secure the assistance of another belayer to back you up if you have any uncertainty regarding the weight differential.
9. Make sure the belay rope is properly positioned on the element and that the participant is clipped in on the proper side for ascending elements.
10. Encourage climbers as appropriate. Ask climbers to indicate the type of support they want from the group — strong verbal encouragement; silence?

Body Belays and Mechanical Belay Devices

A properly executed body belay provides a direct connection between the climber and belayer that can enhance the trust building goals of Project Adventure programs. The use of belay devices allows some belayers to feel more confident, particularly for lighter persons belaying heavier participants. As a general recommendation, we advise practitioners to develop competence with both the body belay and at least one system utilizing mechanical belay devices.

When using a body belay:

1. Run the belay rope through a carabiner on the lower side of the swiss seat or harness on the side opposite to the brake hand.
2. Keep the belay rope under the belayer's butt. When braking, put and keep brake hand down (to crotch).
3. Stand in a position which distributes weight equally with knees slightly flexed, feet spaced slightly wider than shoulders, and has neither foot forward of the other.
4. Wear gloves that fit properly and are suitable for belaying.
5. The belayer should be properly dressed; avoid wearing lightweight clothing, and avoid clothing which is made of synthetic materials. (Exercise clothing made of acrylic, polyester, polypropylene etc., should be avoided. The friction produced in belaying can quickly melt through these materials.) Use extra protective clothing or belay chaps when necessary.

Mechanical Belay Devices

There are many acceptable belay devices in use today. For simplicity and consistency, Project Adventure currently teaches the use of the Figure 8 and the Sticht Plate belay devices.

Hand movement principles are basically the same here as with the body belay. The Figure 8 or Sticht Plate must be securely and properly attached to the front of the swiss seat, Studebaker Wrap, or harness, with a locking carabiner.

When using a mechanical belay device:

1. Pull in the belay rope with both hands; extend braking hand towards climber approximately 12"–14".
2. Holding the rope with the braking hand, slide guide hand above brake hand.
3. Bring hands together.
4. Hold both parts of rope with guide hand, slide braking hand toward body, keeping it ready in case of fall. NEVER take braking hand off the rope.
5. Holding a fall: Smooth movement of the brake hand to the side will increase the friction on the rope and stop the fall.
6. When backup belays are used, the backup belayer can simply stand near the primary belayer and hold the belay rope. When a belay rope is properly run through a Figure 8 or Sticht plate, very little force is required to stop an normal fall.
7. Some Figure 8's have light weight construction and are designed only for rappelling. Do not use these for belaying.
8. If your program uses commercially made harnesses, follow the manufacturer's instructions for the proper connection point when using mechanical belay devices.
9. The use of gloves is not required when using a Figure 8 or Sticht plate belay. Please note that belays here can also sometimes get hot and a close fitting glove is sometimes advisable. *Poorly fitting gloves, as well as other loose clothing and long hair, can get caught in belay devices if proper care is not exercised.*

Common Misunderstandings in Belaying

The following list presents some commonly held misunderstandings about belaying.

A belay system does not give, it is a fixed point.

Not necessarily. There is a significant amount of dynamic built into the belay systems on courses constructed in trees. Size, type of tree, and length of the belay cable determine this dynamic. Any climb which begins below ten feet,

unless the belay point comes from a bolt affixed directly to a limb or solid section of the trunk, will have significant give to it due to the dynamic of the trees and stretch of the rope. On elements such as the Pamper Pole, Vertical Play Pen, Dangle Duo, and Inclined Log, where participants initiate climbing immediately, a fall from a height of less than ten feet will, in all likelihood, bring the climber to the ground. Be aware of this, adjust the belayer's position, and provide spotters if necessary.

The tighter the belay, the more secure it is.

Not always true. Providing too much tension on elements like the Dangle Duo, Heebie Jeebie, Pamper Pole, Vertical Play Pen, Inclined Log, Multivine Traverse, and Cat Walk, can actually pull a person off the element. Allowing the participant a voice in determining the amount of slack that appropriately challenges them, while remaining within safe parameters, is proper.

Keeping a tight belay rope helps the pulley move across the cable.

Not necessarily. If you exert a significant amount of pressure on the belay rope, the pulley can't turn and will impede smooth movement. Too much pressure twists the cheeks of the pulley into the cable, and while the participant can move, it will be in jerky motions — engendering an even more precarious feeling. Tension will probably be wanted at first by participants, but adjust for this as they progress on the element. This is true for all traversing belays.

If a person falls, stopping the fall immediately is the best course of action.

In certain cases (where the dynamics of a belay system — stretch of rope, reaction to fall, give of trees), stopping the fall will actually cause the participant to swing into the element. Examples of this are the Cat Walk, Two Line Bridge, and Tired Two Line. Do not let the participant free fall, but, keeping some pressure on the belay rope, let them descend below the element (in control) and then stop them.

Lowering people down is the same for each element.

Not Necessarily. If this is not done carefully, a person will scrape across the foot cable or log that they were on. Be sure that when lowering a participant off of certain elements, you have the person grab the belay rope, and lean back so that they will comfortably clear the log or cable as they descend. Always use a slow, controlled, and smooth descent.

The belay rope is a good security point for the participant to grab onto.

It depends. Always be aware of which portion of the belay rope the participant chooses to hold on to — the ascending portion or the descending portion. An apprehensive individual may grab the ascending portion when reaching for security and get a fairly painful rope burn.

The backup belayer watches the climber.

Not necessarily. The responsibility of the backup belayer is to provide a fail-safe system to the prime belayer. In this role the backup belayer does not assist the prime belayer in bringing up or letting out rope and should not impede the prime belayer in this role. The backup belayer's attention should be riveted to the prime belayer and he/she should be standing in a position to become the primary belayer in the event that the primary belayer looses control. Therefore, the backup belayer needs to be positioned to always monitor the prime belayer and not the climber. If the backup belayer is backing up a body belay, he/she needs to be skilled in knowing when to intervene.

Backup belayers make the system safer.

Not necessarily. Backup belayers who are not alert and who have not been thoroughly trained may overreact and cause the brake hand of the prime belayer to be pulled away from his/her side. If this is done to a primary belayer using a body belay, that belayer may be pulled or twisted into the backup belayer, causing an unsafe belay situation.

The free end of the belay rope will take care of itself.

Not so. The free end of the belay rope needs to be monitored by the belayer to keep it from hanging up on twigs, stumps, or becoming wrapped or entangled in the belayers feet or legs. Assigning this task to a group member gives the advantage of involving more than just the climber and belayer in the task.

The tightness in the belay rope should always be the same.

Not so. Certain elements require a give and take of rope as participants move up and down — examples: Dangle Duo, Vertical Play Pen, and tree climbs. Belayers should always be aware of this — it helps the participant to not have to fight against the belayer.

Terrain does not effect the belay.

When terrain slopes up or down belayers need to adjust and sometimes give rope when walking down hill, or take rope in as they walk uphill. Look at the activity beforehand and be prepared.

Belayers need not be anchored.

Not necessarily. Certain belayers feel more comfortable anchored when belaying a large person on a fall or jump, and love having terra firma under their feet at all times. Using other group members to hold onto a belayer's swiss seat, Studebaker Wrap, or harness, can help secure the belayer. Giving support on the belayer's shoulders is another method. However, when the belayer is following the movement of a participant on a traversing element, then the support crew needs to move with the belayer and not hinder or interfere with the belayer's hands or the free end of the belay rope. Securing or anchoring the belayer to a fixed point is not done on any traversing belay because of the resulting pendulum fall which could occur.

Errors in belaying technique are infrequent.

Unfortunately not true. Here are some of the more common errors in belaying:

1. Letting go with the brake hand: Allowing the brake hand to lose contact with the belay rope while belaying can happen with both body belays and belay devices. This form of poor technique afflicts even experienced belayers and should be avoided.
2. Improperly extending the arms: Many new belayers transfer the belay rope from brake hand to the fingers of the guide hand with their arms fully or near fully extended. Proper technique requires that the belayer be able to brake a fall at any moment, and this is made difficult by fully extending the arms.
3. Unlocked carabiner: The use of Sticht Plates, Figure 8's, and other similar belay devices should be used with locking carabiners. Always check to be sure that the carabiner gate is locked.
4. Climber inadvertently goes off belay while switching belays: This happens more frequently than commonly recognized. If you decide to have participants switch belays at height, these switches must be very carefully monitored.

Guidelines for Using Participant Belayers

Programs which use participant belayers should follow these steps in training participant belayers:

1. Participants should be thoroughly trained in proper belaying procedures.
2. Back-up belayers should be used.
3. Participant belayers must be carefully supervised by a program instructor.
4. Participant belayers who are not backed-up should have the same qualified level of training and demonstrated competence as program staff.

The High Course Check List

The high elements on a ropes course are usually the most dramatic and are therefore perceived as being the most dangerous. While safety studies tend to disprove this point in terms of frequency of accidents, they do not eliminate the fact that when you are holding a person at a height, where abandonment of the safety system could lead to a serious or even fatal accident, you are totally responsible for that person's physical safety. It is imperative that ropes course instructors be diligent in preparing to lead individuals on high ropes course elements. It is necessary for each and every instructor to go through a series of steps and procedures before taking a group onto a course, and it is

necessary that these be done before each activity. We call these steps and procedures:

Pre-Flight Checks

A. Visually inspect the activity and the area which the activity encompasses for any unnatural or potentially dangerous obstacles. These could include — glass, exposed roots, stumps, rocks, or dead branches on the ground. Make sure the ground offers safe and firm footing. Make sure that there are no dead branches (widow makers) hanging above or on trees where an activity will take place. Give the element and support tree or pole a visual check.

B. Carefully and fully inspect all ropes, carabiners, swiss seats, harnesses, helmets, belay devices, and lead climbing protection devices.

C. Remind all participants of their responsibilities when participating on high elements. These include:

- Never step on any ropes.
- Correctly put on a helmets, and use them when appropriate. (See appendix B).
- Review and check harness systems, swiss seats, or Studebaker Wraps for proper buckling and tying of knots.

D. Be prepared to preform a rescue on the ropes course by having present any extra ropes, a ladder, and hardware which would be needed. This rescue equipment should always be present.

E. Make sure that all belay ropes and hardware are properly set up. This means:

- Belay apparatus (Rosa Pulleys, Spin-Static Pulleys, carabiners, rapid links, and shear reduction blocks) are properly affixed to belay cables and gates are oriented in the correct position and locked.
- Belay ropes are properly oriented for both the belayer and climber.

F. Make sure that belay ropes are not twisted or rubbing anywhere and that only the required knots are tied in the ropes.

G. Be in a position to see and communicate with the participant throughout the activity. (Sun should not be in the belayer's eyes.)

Attachment of Belay Rope to Participant (Climber)

Attaching the belay rope to the participant before he or she ascends into the heights is one of the more important tasks which the instructor performs or observes. New instructors, and perhaps those with a lot of experience, when confronted by mountaineers and rock climbers, may be questioned as to why they clip a carabiner into the front of a swiss seat, Studebaker Wrap, or harness (the instructor may even be told that this method is wrong). Instructors should understand that there are many tie-in systems. Project Adventure, however, is not going to get into the debate as to which is *the only*

right and correct system. We recognize that there are many safe techniques and certain preferences — aka different strokes for different folks. Here are the methods that we teach and endorse:

1. The two knots we allow in the end of the belay rope are the bowline on a bight and the figure 8 loop.
2. Either of these knots, when used in conjunction with a properly tied safety knot, can be used to clip a locking carabiner (or 2 carabiners in the case of the Pamper Pole or Pamper Plank) into a swiss seat, Studebaker Wrap, or harness.
3. When using a swiss seat or Studebaker Wrap, the locking carabiner must be clipped around the overhand knot and all remaining strands of rope which are in front and lying above the participants belly button, or with a rear clip-in, the same strands and knot on the back of a participant and above the glutes.
4. When using a harness, the locking carabiner(s) must be clipped into the portion of the harness designed to accept a carabiner, not necessarily the waist band
5. Once the carabiner(s) is clipped in, it must be locked; i.e., the screw gate must be shut.
6. Project Adventure allows the use of either a front or rear clip-in when participants are tied into a swiss seat or Studebaker wrap. The Studebaker Wrap is preferred for a rear clip-in since it provides for the equal distribution of lift in the front and in the back. Careful assessment of the participant should take place before choosing a rear clip-in.
7. Tying directly into a swiss seat, Studebaker Wrap, or harness is permitted if a figure 8 follow through is used. If this is done, then the belay rope shall follow the exact same set up for clipping a carabiner into the swiss seat, Studebaker Wrap, and harness.

Setting up the High Ropes Course

Instructors Procedures

Setting up the high ropes course must periodically be done by the ropes course instructors. The height and difficulty in climbing to high elements should not exceed the instructors ability and confidence. Consistent practice can improve confidence and increase one's competence. But this should always be done in a controlled environment — practice setting up techniques with a fairly loose overhead belay.

Constant monitoring of belay apparatus and ropes by instructors is paramount to running safe programs irregardless of the system employed.

Project Adventure teaches three methods for setting up belay apparatus and ropes on a high ropes course. All three systems have different advantages and disadvantages and require different levels of expertise before being employed.

1. *Lead climb with partner belaying.* Lead climbing a tree or pole requires two people — a belayer and a climber. The climber attaches the belay rope to him/herself and then leads out (climbs) to the first protection point, usually a Shoulder Lag Eye Screw or 1/2" obliquely placed staple. At the first protection point, a carabiner is clipped-in and the belay rope is clipped into the carabiner so that it runs cleanly through (not twisting the carabiner or having the rope wrap around the carabiner). Eyebolts, Strandvise bales, and looped cable connectors, can also be used if they are properly oriented. Protection points should be placed no more than eight to ten feet apart. The climber ascends, clipping into each subsequent piece of protection, until he/she reaches the belay cable. At this point the climber needs to clip-off (protect oneself) so that the belay rope can be pulled up and reeved through the belay device affixed to the belay cable. Once this is completed, the climber climbs down on belay, removing carabiners left at the protection points.

2. *Pull belay rope up with haul cord.* Haul cords are usually made of No. 4 nylon (sometimes referred to as parachute cord). These are attached (tied) into eye screws melted into the end of belay ropes. Once tied in, the belay ropes are pulled up through the belay apparatus. Reverse the procedure to remove the belay rope.

3. *Use an adjustable self-belay (Web's Wings) for lead climbing and protection.* An adjustable self-belay allows instructors to climb a tree or pole and be protected throughout the climb. The adjustability of each wing not only allows instructors to clip-off easily at the belay cable, but provides an extra wing to clip onto the belay cable so that the belay apparatus will not slide away when affixing it to the belay cable.

   An instructor climbs using an adjustable self-belay by alternately clipping each wing into a protection point while ascending. After clipping in the top wing, the climber down climbs, resting his/her weight on the top wing for balance or rest, and unclips the bottom wing which will then be moved up as the climber ascends, to become the top wing. Upon reaching the belay cable, the instructor can adjust one of the wings to go around the tree or pole and clip it back into the swiss seat, Studebaker Wrap, or harness to provide a sound anchor.

High Course Elements

The high ropes course elements listed here are called by the names used by Project Adventure; in parentheses are other names by which these elements have been known:

Inclined Log

Two Line Bridge (Postman's Walk)

Cat Walk (High Balance Log)

Heebie Jeebie (High Criss Cross, Hourglass)

Burma Bridge (Eagle Walk, Three Line Bridge)

Tired Two Line

Dangle Do (Dangle Duo, Log Ladder)

Multivine Traverse

Vertical Play Pen

Centipede

High Tension Traverse

Pamper Pole (Trapeze Dive)

Pamper Plank (Trapeze Dive)

Zip Wire (Flying Fox)

This list is representative of elements that Project Adventure regularly builds. On occasion, the Project builds other elements and many programs have additional elements other than those listed. For the most part; the principles of belaying, whether a ground belay or static belay, remain constant. However, if other elements are built, it is recommended that additional pages be added to this manual to cover those activities, especially if this manual is to be used as a guide for instructors at your program.

Inclined Log

A substantially sized peeled log or pole approximately thirty to thirty-five feet in length, which is attached to a tree at an incline of not more than 20 degrees.

Task:

For an individual, clipped into an overhead belay, to walk, shimmy, or crawl to the uppermost part of the log or pole.

Instructor's Role:

A. Complete the pre-flight check.

B. Correctly tie the proper knots into the end of the belay rope.

C. Clip a locking carabiner into the swiss seat, Studebaker Wrap, or harness, and double check that the gate is clipped around all the proper strands and screwed shut.

D. Before climbing, check and make sure that the swiss seat, Studebaker Wrap, or harness is tied correctly with safety knots, and make sure that the knots are not loose.

E. Make sure that the participant is wearing a correctly sized helmet and that it is put on according to the manufacturer's instructions.

F. Make sure that there are two to four spotters ready to accompany the participant up the log until the participant is six feet above the ground.

G. Set up the element in such a way that the belayer can move freely with the traversing participant. The belay apparatus should always remain above the climber.

H. Do not have the belayer stand so far away from the element that if a participant falls, the belayer will be pulled towards the element and not be able to arrest the fall.

I. If possible, have the belayer stand in a position where a falling participant will not be able to grab the part of the belay rope that runs to the belayer.

Comments or Special Considerations:

Two Line Bridge

Two parallel cables strung horizontally between trees or poles and approximately four to four and a half feet apart. A third cable is often oriented above the second cable and serves as the belay cable. Where a third cable is not present, the top cable serves as the belay cable/hand cable.

Task:

To climb to the element and traverse across with the lower cable serving as the foot cable, and the upper cable providing support for the hands.

Instructor's Role:

A. Complete the pre-flight check.

B. Correctly tie the proper knots into the end of the belay rope.

C. Clip a locking carabiner into the swiss seat, Studebaker Wrap, or harness, and double check that the gate is clipped around all the proper strands and screwed shut.

D. Before climbing, check and make sure that the swiss seat, Studebaker Wrap, or harness is tied correctly with safety knots, and make sure that they are not loose.

E. Set up the element in such a way that the belayer can move freely with the traversing participant. The belay apparatus should always remain above the climber.

F. Do not have the belayer stand so far away from the element that if a participant falls, the belayer will be pulled towards the element and not be able to arrest the fall.

G. If possible, have the belayer stand in a position where a falling participant will not be able to grab the part of the belay rope that runs to the belayer.

Comments or Special Considerations:

Cat Walk

A horizontally positioned log or pole suspended between two trees or poles. The belay cable is positioned above the log, parallel to the ground, and at a height of nine to ten feet above the log.

Task:

To climb to and traverse the log.

Instructor's Role:

A. Complete the pre-flight check.

B. Correctly tie the proper knots into the end of the belay rope.

C. Clip a locking carabiner into the swiss seat, Studebaker Wrap, or harness, and double check that the gate is clipped around all the proper strands and screwed shut.

D. Make sure that the participant is wearing a correctly sized helmet and that it is put on according to the manufacturer's instructions.

E. Before climbing, check and make sure that the swiss seat, Studebaker Wrap, or harness is tied correctly with safety knots, and make sure that they are not loose.

F. Set up the element in such a way that the belayer can move freely with the traversing participant. The belay apparatus should always remain above the climber.

G. Do not have the belayer stand so far away from the element that if a participant falls, the belayer will be pulled towards the element and not be able to arrest the fall.

H. If possible, have the belayer stand in a position where a falling participant will not be able to grab the part of the belay rope that runs to the belayer.

Comments or Special Considerations:

Heebie Jeebie

Two diagonally crossing multiline ropes are connected from support trees or poles to the taut traverse foot cable, at points two thirds the distance from the trees or poles. A belay cable runs approximately nine to ten feet above and parallel to the foot cable.

Task:

To climb to and traverse the activity while walking between or on the outside of the crossing ropes.

Instructor's Role:

A. Complete the pre-flight check.

B. Correctly tie the proper knots into the end of the belay rope.

C. Clip a locking carabiner into the swiss seat, Studebaker Wrap, or harness, and double check that the gate is clipped around all the proper strands and screwed shut.

D. Before climbing, check and make sure that the swiss seat, Studebaker Wrap, or harness is tied correctly with safety knots, and make sure that they are not loose.

E. Set up the element in such a way that the belayer can move freely with the traversing participant. The belay apparatus should always remain above the climber.

F. Do not have the belayer stand so far away from the element that if a participant falls, the belayer will be pulled towards the element and not be able to arrest the fall.

G. If possible, have the belayer stand in a position where a falling participant will not be able to grab the part of the belay rope that runs to the belayer.

H. Recognize that falls can occur when a participant is near the center. If a fall occurs, arrest the fall so that the participant does not get a foot caught between the diagonal ropes and the foot cable.

I. Anticipate that if a person falls and is belayed to the ground, the belay rope may not follow a proper lead for the next climber and correct for this.

Comments or Special Considerations:

Burma Bridge

A V-shaped bridge is formed between two trees or poles with the apex of the V serving as the foot cable and the two tops of the V serving as hand rails. An overhead belay cable, parallel to the ground and out of a participant's reach, completes the activity. Sometimes this element is accompanied by a series of ropes connected from the hand cable to give some lateral stability or to make the element visually more aesthetic.

Task:

To walk across the foot cable while using the two hand cables for support.

Instructor's Role:

A. Complete the pre-flight check.

B. Correctly tie the proper knots into the end of the belay rope.

C. Clip a locking carabiner into the swiss seat, Studebaker Wrap, or harness, and double check that the gate is clipped around all the proper strands and screwed shut.

D. Before climbing, check and make sure that the swiss seat, Studebaker Wrap, or harness is tied correctly with safety knots, and make sure that they are not loose.

E. Set up the element in such a way that the belayer can move freely with the traversing participant. The belay apparatus should always remain above the climber.

F. Do not have the belayer stand so far away from the element that if a participant falls, the belayer will be pulled towards the element and not be able to arrest the fall.

G. If possible, have the belayer stand in a position where a falling participant will not be able to grab the part of the belay rope that runs to the belayer.

H. Anticipate that if a participant falls, or is belayed down in the middle of the activity (and you have rope supports between the hand rails), you now face the situation of having a belay rope stuck in the middle and must correct this. Do not belay an individual up with the belay rope in this position. Reposition the belay rope to allow a participant to climb straight up to the element.

Comments or Special Considerations:

Tired Two Line

This element is similar to the Two Line Bridge but without the hand cable. It does provide two foot cables (one of which was the hand cable that needed a rest). The distance between the cables varies from one to two feet. They are usually attached to turnbuckles to allow for changes in cable tightness — making it harder when they are loose and easier when they are taut. A belay cable runs above and out of reach of the participant and parallel to the ground.

Task:

To traverse the activity with one foot on each cable.

Instructor's Role:

A. Make sure the hook portion of the turn-buckle is properly clipped into the eye bolt and strandvise in the tree or pole. The tip of the hook should be oriented towards the ground. *Note*: Foot cables need to be tight enough to prevent turnbuckles from unhooking themselves from the eyebolt or strandvise.

B. Complete the pre-flight check.

C. Correctly tie the proper knots into the end of the belay rope.

D. Clip a locking carabiner into the swiss seat, Studebaker Wrap, or harness, and double check that the gate is clipped around all the proper strands and screwed shut. *Note*: If a rear clip-in is used, the belayer must be able to arrest a fall before a participant falls forward and has a chance of striking his/her head against the cables.

E. Before climbing, check and make sure that the swiss seat, Studebaker Wrap, or harness is tied correctly with safety knots, and make sure that they are not loose.

F. Set up the element in such a way that the belayer can move freely with the traversing participant. The belay apparatus should always remain above the climber.

G. Do not have the belayer stand so far away from the element that if a participant falls, the belayer will be pulled towards the element and not be able to arrest the fall.

H. If possible, have the belayer stand in a position where a falling participant will not be able to grab the part of the belay rope that runs to the belayer.

Comments or Special Considerations:

Dangle Do

A vertically oriented log ladder suspended from an overhead cable or clipped directly into support trees or poles. The logs are usually 8' pressure treated landscaping timbers. A separate belay cable is suspended above the uppermost log.

Task:

A participant or participants (2, since this activity is often done in pairs) has to climb the ladder using only the support of the logs and/or the other participant. Use of the side cables for climbing is discouraged.

Instructor's Role:

A. Complete the pre-flight check.

B. Correctly tie the proper knots into the end of the belay rope.

C. Clip a locking carabiner into the swiss seat, Studebaker Wrap, or harness, and double check that the gate is clipped around all the proper strands and screwed shut.

D. Before climbing, check and make sure that the swiss seat, Studebaker Wrap, or harness is tied correctly with safety knots, and make sure that they are not loose.

E. Make sure the participants are wearing a correctly sized helmet put on according to the manufacturer's instructions.

F. When mounting the first log, have two spotters spot each participant (especially if its fairly low to the ground). Rope stretch at this point will prevent a belayer from arresting a fall.

G. Do not have the belayer(s) stand so far away from the element that if a participant falls, the belayer will be pulled towards the element and not be able to arrest the fall.

H. If possible, have the belayer(s) stand in a position where a falling participant will not be able to grab the part of the belay rope that runs to the belayer.

I. Do not allow participants to wrap their belay rope around horizontally suspended logs on their upwards journey.

J. Have another participant pull the Dangle Do out of the way when belaying a participant or participants down.

Comments or Special Considerations:

Multivine Traverse

A single, tensioned foot cable with a series of multiline ropes suspended from an overhead cable, positioned just beyond the average persons' reach.

Task:

To walk across the foot cable using the various support vines for aid.

Instructor's Role:

A. Complete the pre-flight check.

B. Correctly tie the proper knots into the end of the belay rope.

C. Clip a locking carabiner into the swiss seat, Studebaker Wrap, or harness, and double check that the gate is clipped around all the proper strands and screwed shut.

D. Before climbing, check and make sure that the swiss seat, Studebaker Wrap, or harness is tied correctly with safety knots, and make sure that they are not loose.

E. Set up the element in such a way that the belayer can move freely with the traversing participant. The belay apparatus should always remain above the climber.

F. Do not have the belayer stand so far away from the element that if a participant falls, the belayer will be pulled towards the element and not be able to arrest the fall.

G. If possible, have the belayer stand in a position where a falling participant will not be able to grab the part of the belay rope that runs to the belayer.

Comments or Special Considerations:

Vertical Play Pen

Similar to the Dangle Do but providing for a pot pourri of challenges on the upwards climb. Some of these include mounting and balancing on a regain rope, climbing a short etrier, a short rope ladder, mantling over a large truck tire, and crawling up through another truck tire. This is usually a single participant activity but can be done with two participants.

Task:

To climb upwards over the series of presented obstacles using your balance, dexterity, and creativity in surmounting each task.

Instructor's Role:

A. Complete the pre-flight check.

B. Correctly tie the proper knots into the end of the belay rope.

C. Clip a locking carabiner into the swiss seat, Studebaker Wrap, or harness, and double check that the gate is clipped around all the proper strands and screwed shut.

D. Before climbing, check and make sure that the swiss seat, Studebaker or harness is tied correctly with safety knots, and make sure that they are not loose.

E. Do not have the belayer stand so far away from the element that if a participant falls, the belayer will be pulled towards the element and not be able to arrest the fall.

F. If possible, have the belayer stand in a position where a falling participant will not be able to grab the part of the belay rope that runs to the belayer.

G. Make sure the participant is wearing a correctly sized helmet put on according to the manufacturer's instructions.

H. When mounting the first obstacle, have two spotters spot each participant (especially if they are fairly low to the ground). Rope stretch at this point will prevent a belayer from arresting a fall.

I. Do not have the belayer stand so far away from the element that if a person falls, the belayer will be pulled towards the element and not be able to arrest the fall.

J. Do not allow participants to wrap their belay rope around horizontally suspended logs, ropes, or tires, on their upwards journey.

K. Have another participant pull the lower section of the Vertical Play Pen out of the way when belaying a participant or participants down.

Comments or Special Considerations:

Centipede

A series of vertically suspended 4'x4's which have randomly placed 1/2" staples to provide foot and hand holds. A separate belay is located on either a belay cable or bolt.

Task:

To climb the undulating 4'x4's using the staples as points of contact for feet and hands.

Instructor's Role:

A. Complete the pre-flight check.

B. Correctly tie the proper knots into the end of the belay rope.

C. Clip a locking carabiner into the swiss seat, Studebaker Wrap, or harness, and double check that the gate is clipped around all the proper strands and screwed shut.

D. Before climbing, check and make sure that the swiss seat, Studebaker Wrap, or harness is tied correctly with safety knots, and make sure that they are not loose.

E. Do not have the belayer stand so far away from the element that if a participant falls, the belayer will be pulled towards the element and not be able to arrest the fall.

F. If possible, have the belayer stand in a position where a falling participant will not be able to grab the part of the belay rope that runs to the belayer.

G. Have a participant pull the Centipede off to one side and away from the participant being belayed down.

Comments or Special Considerations:

High Tension Traverse

Exactly the same as a Low Tension Traverse except that this activity is belayed from a separate cable 10–12 feet above the foot cable.

Task:

To take skills perfected and developed on the Tension Traverse at ground level and test those skills at height.

Instructor's Role:

A. Complete the pre-flight check.

B. Correctly tie the proper knots into the end of the belay rope.

C. Clip a locking carabiner into the swiss seat, Studebaker Wrap, or harness, and double check that the gate is clipped around all the proper strands and screwed shut.

D. Before climbing, check and make sure that the swiss seat, Studebaker Wrap, or harness is tied correctly with safety knots, and make sure that they are not loose.

E. Make sure that the participant wears a correctly sized helmet, put on according to the manufacturer's instructions.

F. Set up the element in such a way that the belayer can move easily with the traversing participant. The belay apparatus should always remain above the climber.

G. Participants should be instructed to let go of the Tension Traverse if they fall.

H. The belayer must be positioned and alert to the possibility of a pendulum fall.

I. Do not have the belayer stand so far away from the element that if a participant falls, the belayer will be pulled towards the element and not be able to arrest the fall.

J. If possible, have the belayer stand in a position where a falling participant will not be able to grab the part of the belay rope that runs to the belayer.

Comments or Special Considerations:

Pamper Pole & Pamper Plank

Task:

These activities involve climbing to the top of a pole, or to a platform, and diving out to a trapeze suspended from a cable. The distance to the trapeze varies according to the needs of a particular program, but distances generally range from five to nine feet. The primary difference between the Pamper Pole and Pamper Plank is that when a participant climbs to the top of a Pole, they must come to a standing position and balance themselves on the top; whereas on the Plank, they must walk out to the end of the Plank before jumping for the trapeze. In both cases a separate belay cable is suspended above the trapeze cable. The belay rope should run through a Shear Reduction Block or Spin Static Pulley affixed to the belay cable.

Instructor's Role:

A. Complete the pre-flight check.

B. Correctly tie the proper knots into the end of the belay rope.

C. Clip **two** locking carabiners into the swiss seat, Studebaker Wrap, or harness. Double check that the gates are clipped around all the proper strands and screwed shut. Note: If a commercial harness or tied chest harness is used, see appendix B.

D. Before climbing, check and make sure that the swiss seat, Studebaker Wrap, or harness is tied correctly with safety knots, and make sure that they are not loose.

E. Make sure that the participant wears a correctly sized helmet, put on according to the manufacturer's instructions.

F. Set up the element to have the belay rope run through the Just-Rite Descender — as the participant jumps, the belayer moves backwards to take up slack.

G. Make sure that if you are not using the Just-Rite Descender the belayer does not stand so far away from the element that if a person falls, the belayer would be pulled towards the element and not be able to arrest the fall.

H. Have the belayer stand in a position where a falling person will not be able to grab the part of the belay rope that runs to the belayer.

Comments or Special Considerations:

Zip Line

A long cable attached to a tree, pole, or tower, with a platform situated six to eight feet below the cable, descends to a stationary anchor point (tree, pole, building). The descent is accomplished by a participant being attached to a properly designed pulley which rides on the cable.

Task:

For a participant to climb to the platform while on belay, and then be properly clipped into a pulley which allows the participant to ride the pulley down the cable to a point where he/she is stopped by either a gravity brake system or bungee brake.

Instructor's Role:

A. Complete the pre-flight check.

B. Correctly tie the proper knots into the end of the belay rope.

C. Clip a locking carabiner into the swiss seat, Studebaker Wrap, or harness, and double check that the gate is clipped around all the proper strands and screwed shut.

D. This activity usually requires switching from a ground belay to a direct clip in to the Two Wheel (Zip) Pulley. The instructor must carefully monitor this switching of belays, which usually means that the instructor is on the platform switching belays.

E. Instructors on the platform should be clipped in or tied off for their own safety.

F. The Instructor (usually) must correctly attach the participant to the direct clip in on the Zip Pulley before the ground belay is removed.

G. Remove any self-belay or static system that was used to secure the participant before letting them descend.

H. Provide a secure means for unclipping participants from the pulley and descending to the ground after the ride is complete. This may involve having a step ladder available at the end of the ride with other members of the group available to assist and spot the rider down.

I. Make sure that if a bungee brake system is employed, proper training has taken place before the activity is run.

Comments or Special Considerations:

Conclusion:

This manual is designed to aid those in the field running Adventure Programs which utilize a Challenge Ropes Course. While it is geared to focus specifically on Project Adventure Programs, it is also supported by other site specific manuals from respected programs.

The goal is not to certify the correct standards for all Rope Course Programs but to assure Adventure Educators that there is a genuine body of information available as a resource for the field.

This manual is not a text on how to run a program and will not come close to replacing knowledge gained at professional training seminars, but it will supplement that information.

Appendix A

Program Accreditation

For Whom?

Program Accreditation is designed primarily for those programs that are within the Project Adventure network; that is, programs that have received PA training, used publications and curriculum materials published by PA, or received other PA services such as ropes course construction. Programs that have had little or no contact with PA are, however, welcome to seek accreditation if they wish.

Why?

Programs that undertake the accreditation process are seeking outside evaluation of their programs with regard to quality and safety. The term accreditation means "formal written confirmation"; these programs are seeking "confirmation" that their programs are within the current standards of safety.

This assurance may be useful for making changes in program equipment and/or design, and in providing information on program quality to third parties, such as administrations, insurance companies, and the public.

The Accreditation Process

1. Programs that wish accreditation should ask for an application form from either the Massachusetts or the Georgia office. After a completed application is received, the scheduling and planning of the particular visit can be accomplished.
2. The Program Accreditation involves the following components:
   a) on-site program observation (usually one day with an additional day for ropes course safety inspection) — one or two observers
   b) observation of program sessions
   c) ropes course safety inspection, including the inspection of all safety equipment — ropes, slings, harnesses, etc.
   d) review of program written materials; e.g., curriculum, program goals and objectives, safety and program procedures, evaluation materials
   e) review of all high and low ropes course procedures with staff. (This takes place on the ropes course and involves demonstration as well as discussion.)
   f) written evaluation to be completed by each staff member

3. A written report will be sent following accreditation. This report will include an appraisal of staff skills and recommendations for further training, if necessary, and for program design options to consider. Just as the accreditation process itself is a customized approach (while maintaining standards), the report can be customized to meet the needs of the institution.

Appendix B

Equipment Standards Recommended by Project Adventure, Inc.

Ropes

We use two kinds of rope for belaying and swiss seats; 1.) kernmantle rope — kern (core), mantle (sheath), and 2.) hawser-lay rope (three strand twisted).

Why two kinds? We like the handling characteristics of kernmantle rope; it holds a knot well, functions extremely well with a large Figure 8 belay, has less stretch under normal loading, and passes U.I.A.A. certification. The two sizes most often used are 11mm and 9mm: 11mm is used for belaying, and 9mm is used for swiss seats.

The three strand hawser-lay rope is less expensive, has more resistance to abrasion, can be examined more easily for wear, and exhibits more stretch under load. Many programs prefer hawser-lay rope for these reasons. We sometimes suggest this rope for Pamper Pole/Plank and Trapeze jumps — when you want greater rope stretch for a more dynamic belay. But remember, if you have a short Pamper Pole/Plank or Trapeze Jump, this increased stretch can work against you if you have a heavy jumper on the low pole. Use kernmantle rope for jumps under 25 feet.

Carabiners

Karl Rohnke has a collection of over 75 carabiners — in all different shapes and sizes, which is a sure bet that there are many more. Project Adventure uses different types of carabiners for different applications. These are: Locking, Non-Locking, Dog Leg, and Twist lock.

Locking Carabiners:

Currently we use three types of locking carabiners; Clog 11mm locking D-aluminum, Clog 11mm locking D-steel, and Stubai locking oval-steel. Locking carabiners are used in all belay clip-in situations, occasionally on belay cables, and with ROSA Gold Pulley (see Belay and Pulley Systems).

Non-Locking Carabiners:

The most common use of non-locking carabiners is for setting up elements such as the Spiders Web, clipping in swing ropes and Tension Traverse ropes, and in use on a swiss seat or Studebaker Wrap to hold the belay rope under the belayer's butt (body belay). Currently we use the 11mm Clog.

The use of non-locking carabiners is not acceptable in any belay system.

Dog Leg Carabiners:

These should be used for clipping-in ropes to swings, putting up Spider Webs, etc. They should not be used for belays (on cable), clip-in points on swiss seats, or on climbing leads (Webs Wings, Lobster Claws, etc.)

Twistlock Carabiners:

These are self-locking carabiners that automatically lock when the gate closes. While they can be used in place of a non-locking carabiner, their cost is usually prohibitive. They are the best choice to use with Web's Wings or Bear Claws as they automatically lock you to your protection point when lead climbing.

Always use two locking carabiners on the belay for Pamper Pole, Pamper Plank or Trapeze Jump. This is a direct clip-in from the belay rope to the swiss seat, Studebaker Wrap, or harness. When using the chest harness, a third carabiner should be employed from the bowline on a bight to the chest harness.

Belay and Pulley Systems

Carabiners used on belay cables

Pulley systems have replaced carabiners on belay cables for most programs. However, carabiners properly used are acceptable for use on belay cables. When used, we recommend running the belay rope through 2 locking steel carabiners. Carabiner gates should be locked in a downward direction (locking carabiners found open in this application are almost always found to have been locked in an upward direction.) Opposing or reversing the carabiner gates is not necessary in this application. Two steel, 1/2 inch rapid links used in the same way are also acceptable. *Note*: Even steel carabiners and rapid links can wear through when used on steel cable. Ongoing inspection of all devices used on belay cables is essential.

Rapid Links

Also called quick links, these are carabiner like devices that have become popular because of their relatively low cost and very high strength. The breaking strength of a 1/2 inch rapid link with the gate closed exceeds 22,000 lbs. The type of link we have used and tested is called *maillon rapide*. It has its name and country of origin (Made in France) stamped on it and has a safe working load of 1500 kg.

ROSA Gold Pulley

This pulley is used to support belay ropes on elements with traversing belays. The case hardened wheel (sheave) runs on the belay cable. A ROSA Gold pulley should always be used in conjunction with two locking carabiners, or two 1/2" rapid links clipped to the Spin Static belay device. The belay rope should run through the Spin Static pulley or through the carabiners.

ROSA Gold pulley with 2 locking D's

This system is employed where the purchase of a Spin Static pulley is prohibitive. It is also used on activities such as a Two Line Bridge, where the hand cable is the belay cable. Both locking carabiners should be connected to the pulley (not to each other), with the belay rope running through both carabiners. Carabiner gates should be locked in a downward direction.

Spin Static or S.S. Pulley

These should be used in conjunction with a ROSA Gold pulley on a traversing belay. The Spin Static pulley is connected to the Gold pulley with two 1/2" rapid links. Gates should always be locked in a downward direction and connected to each other (in tandem). Using two rapid links properly orients the pulley and allows it to rotate freely.

When the belay is not traversing; e.g., Dangle Duo, connect each Spin Static pulley directly to the belay cable with 2 1/2"rapid links. The Spin Static is a shear reducing pulley which is designed to strengthen the belay system and greatly reduce wear on the belay rope. It is most helpful on activities where participants regularly jump or are lowered down. *In normal belay use, the pulley should be used with both through bolts in place.*

Zip Wire Pulley (Two Wheel)

The ROSA Two Wheel pulley was designed primarily for use as a Zip Wire pulley. It is also used to support a belay rope running through a Spin Static pulley on traversing belays (replacing the ROSA Gold pulley.) When used for a Zip Wire pulley, participants are connected to the pulley with a direct clip-in. The direct clip-in is most commonly a 3 foot section of 5/8" Multiline with an eye splice in each end. This allows the participant's harness or swiss seat to be connected directly to the pulley. The use of a separate backup to this system is not required, although some programs prefer to use a backup. If the pulley used is not a ROSA Two Wheel Zip Wire Pulley, then the use of a separate backup is required. (See below.)

Other Zip Wire Pulleys

Other pulleys used for the Zip Wire may be suitable but should be carefully evaluated before being used. If such a pulley is being employed, use a backup (3/8" adjustable multiline prusik, attached to a ROSA Gold pulley and the back of the participant's harness/swiss seat). You should note that the ROSA Gold pulley can jam in this application if the backup is not properly adjusted. Keep the backup fairly short so that there is downward pressure on the Gold pulley.

Shear Reduction Block (SR Block)

This is a larger shear reducing device which is commonly used on activities where participants dive for a trapeze; e.g., a Pamper Pole. It is attached to the belay cable with 1/2" rapid links and has a swaged cable backup that is also attached to the belay cable.

Rescue Pulleys

Pulleys designed for rescue work usually have aluminum wheels (and sometimes nylon) which wear out quickly on steel cables. Construction is usually lightweight with small through bolts and fairly low breaking strength. They are not recommended for ropes course use.

> *Note*: The applications described above refer to ordinary use on outdoor tree and telephone pole ropes courses. Indoor applications of these devices are somewhat different and are not covered in this manual.

Tie-in Systems

Swiss Seat

The term *swiss seat* refers to a harness-like arrangement tied from a 16' to 21' piece of 9 mm rope or 1" webbing. Although there are 4 or 5 commonly used designs, all have both leg loops and a waist band incorporated into a single unit. If properly tied (waist band above the hip bones), a participant cannot fall out of this arrangement. Relative comfort when suspended is achieved by having the participant's weight distributed between the waist and the leg loops. A participant using a swiss seat usually ties or clips into the front of the seat, although a rear clip-in is acceptable for most ropes course uses. *Note*: The rear clip-in used with a regular swiss seat provides somewhat less support than the Studebaker Wrap and safe "hang time" is therefore less. When used, extra care should be exercised to ensure that participants are carefully and quickly brought to the ground and not left hanging.

Studebaker Wrap

A Studebaker Wrap is essentially a double swiss seat made from approximately 26 ft. of 1" tubular webbing or 9 mm climbing rope. The resulting harness has more waist bands and double leg loops on each leg. When properly tied, the Studebaker Wrap does provide some additional support and comfort to the participant. It is suitable for either a front or rear clip-in.

Commercial Climbing Harnesses

Climbing harnesses can be useful for many programs. However, there are many different harness designs and wide variations in their suitability for ropes course use. *It is recommended that you only use a commercial harness in a manner consistent with the manufacturer's instructions for proper use.* For climbing harnesses designed for recreational climbing, this almost always means it should only be used with a proper front clip- or tie-in.

Chest Harnesses

Sometimes we use a commercial or tied chest harness on participants diving from trapeze jumps. We do this because it makes the jumpers more comfortable, is easy to put on, and reduces the likelihood of tumbling falls. A chest harness should never be used without a swiss seat or Studebaker Wrap. The use of a chest harness for a rear clip-in with a Studebaker Wrap gets the rope out of the jumpers face. A chest harness should only be used with thorough and comprehensive training.

Helmets

Each program needs to determine its own policy for helmet use. Project Adventure requires the use of helmets on the following activities:

- All belayed indoor activities
- Pamper Pole, Pamper Plank
- Dangle Do
- Vertical Play Pen
- Climbing Walls and Towers
- Inclined Log
- Cat Walk
- High Tension Traverse

Many programs require helmets on all belayed events for ease of management. This is an acceptable standard.

Helmets should meet U.I.A.A. standards and not be retro fits from your football, hockey, lacrosse, or Easy Rider days.

Knots and Coils

Project Adventure, for the most part, teaches six knots for ropes course application. These are:

- Square Knot
- Double Fisherman's Knot
- Bowline
- Bowline on a Bight
- Figure 8 Loop
- Overhand Knot

Other knots which are useful to know are: Butterfly, Clove Hitch, Prusik, and the most show stopping and esoteric of all knots — the John Wayne Bowline.

Instructors should be well versed in all of the above knots, know when to use them, and be able to teach each on in an understandable fashion.

Proper rope maintenance begins with knowing how to coil a rope. Instructors should know the following coils:

- Mountaineering
- Butterfly (Swiss Coil or Rescue Coil)

Changing Standards

For a variety of reasons activities and procedures change. It is our hope that the newer procedures we use will be helpful and beneficial to the many individuals and organizations with whom we work. It is important to note that not all outdated procedures are inherently unsafe. We recommend that practitioners stay well informed concerning any changes in current practices.

The following is a sample list of some of the activities and procedures that are no longer used by Project Adventure.

Electric Fence — The information gathered in both the 10 and 15 safety studies indicated that this activity was statistically more likely to result in injury. For this reason we no longer use the electric fence in PA training workshops. (see appendix D)

Flea Leap — The traditional Flea Leap involved jumping from one platform (typical height 7 ft.) to another platform (4 to 5 ft.). The difficulty of safely spotting this activity is the major reason its use was discontinued quite a few years ago.

Bowline, bowline on a coil — rope and webbing swami tie-ins. Tying-in with these methods was once widely used but now represents outdated procedures. Current methods involve the use of swiss seats, Studebaker Wrap, and commercial climbing harnesses.

Zip Wire changes — Traditionally, Zip Wire riders held on to hand loops which were connected to an industrial pulley. A bowline around the waist attached to a steel carabiner on the Zip Wire cable served as a backup. Some form of trust brake, which involved other participants armed with implements of deceleration (ropes, tarps, sleeping bags), may have slowed the participant. Current participants ride a 2 wheel pulley designed specifically for Zip Wires. A direct clip-in, usually a 3 ft. piece of 5/8" multiline with an eye splice in each end, connects the pulley directly to the participant's swiss seat, Studebaker Wrap, or harness. Deceleration is now usually accomplished by the combined effects of gravity, a Zip Wire brake block, and a long piece of 1/2"shock cord.

Twist Lock Carabiners — Self-locking carabiners were initially used as a means to insure that the gate would be locked. However, if something rubs up against them, they can open. This often has resulted in connecting a participant to something in addition to their belay rope, such as an element foot cable. Our current recommendation is that twistlock carabiners not be used to connect climbers to their belay. The end "tails" of self-belays; e.g., Bear Claws used by instructional staff to set up the ropes course, is an

acceptable and recommended use of twist lock carabiners (see ropes course set up, pg. 70).

Inclined Belay over the Inclined Log — Participants on the Inclined Log were formerly connected to a piece of goldline which ran parallel to and about 8 ft. above the log. This form of backwards goldline zip has been replaced by a horizontal overhead belay cable.

Standing hip belay without a harness/seat — The current procedure has the belayer wearing a Studebaker, harness, etc., with the belay rope running through a guide carabiner on the harness opposite the belayer's brake hand. This has replaced an older practice of belaying without a harness or guide carabiner.

Stationary Hickory Jump Bar — The original Hickory Jump bar was attached directly to the tree/pole supports. The current practice utilizes a detachable pvc trapeze that provides a reasonably dynamic catching surface and can be removed to limit unauthorized use of the activity.

Single rapid link connected to Spin Static Pulley — For the past several years we have recommended using 2 1/2" rapid links (connected end to end) as means of connecting a ROSA Gold pulley to a Spin Static pulley. This allows the Spin Static pulley to rotate freely reducing drag on the Gold Pulley.

Appendix C

15 Year Safety Study

Project Adventure began as a model program in the Hamilton-Wenham Regional School District in 1971 in Hamilton, Massachusetts. An interdisciplinary approach, with components in physical education, academics, and counseling, the program was granted National Demonstration Site status by the National Diffusion Network of the U.S. Office of Education in 1974, and was widely disseminated to schools and other educational agencies in the years following. Hospitals, camps, counseling centers, colleges, and others, have implemented one or more aspects of the program.

The physical activities of the curricula involve a series of games, initiative problems, service projects and low and high ropes course events. Some programs have camping components or top rope climbing instruction.

In 1981-82, we completed a safety study of the first ten years of implementation of the Project Adventure curriculum. We sent surveys to 246 PA program sites and received back 116, for a return rate of 47%. The accident rate for this survey was a total of 78 accidents, for a total of 15,190,864 hours on task, or a rate of 5.13 accidents per million hours.

This Ten Year Safety Study has provided useful information for many existing programs and for persons seeking to start a Project Adventure program. Project Adventure, Inc., now sees part of its mission as serving as an information center for PA programs. This Fifteen Year Study is an update of our Ten Year Study.

Fifteen Year Survey Returns

In February 1986, in conjunction with the Rhulen Insurance Agency of Monticello, NY, we sent out a total of 725 survey instruments (see exhibit A) to program sites. These sites represented programs with which PA had either trained staff, constructed equipment on-site, or supplied with information, publications, or equipment.

By September of 1986, when we closed the returns, we had a total of 392 programs responding, or a return rate of 54%. The average number of years in operation was 6.5 years. Of the 392 responding, 20 had no ropes course, 34 had low elements only, and 340 had a combination of high (belayed) elements and low (non-belayed) elements.

The breakdown of the respondents by type of institution was as follows:

Schools (Public & Private)	143	37%
Camps	112	28%

Outdoor Education & Nature Centers	64	16%
Colleges	35	9%
Hospitals	24	6%
Municipal Recreation Centers	14	4%
Totals	392	100%

This diversity of sites represents a trend in the field. In 1981, over 80% of the respondents were from schools.

Accident Rate/Definitions

The statistical reporting of injuries is usually done by either a thousand days, a hundred thousand days, or a million hours of exposure to a program. We chose the latter for our Ten Year Study, as it seemed to be easier to compute the "hours" of a standard physical education course, and we have stayed with this measure. On our survey, we asked people to estimate the average length of a participant's program in hours per week. We then used the length of time in use in years, and the number of participants, to arrive at a total participant hours for each site. (Example: A high school physical education class meets for 2 hours per week for 36 weeks would have a total participant hour exposure per year of 80 x 72= 5760.)

We used on this survey the standard National Safety Council and the typical industrial standard of a reportable accident: an injury which results in at least one day lost from school or work.

Using the above definitions, we computed a total of 157 accidents and a total of 42,752,242 hours on task, for a rate of 3.7 accidents per million hours on task.

Comparable Statistics

The injury rate on the Ten Year Study had 5.1 injuries per million hours, compared to the 15 Year rate of 3.7. The difference, we believe, is primarily a result of the tightened definition on the 15 year survey of an accident. In 1981, we received from some programs lists of scrapes and bruises which clearly did not result in any lost time, and, therefore, would not have been reported on the 1986 study. It is also likely, but difficult to prove, that the increasing safety consciousness of the past four years (since the publication of the Ten Year Safety Study) has resulted in more training and attention to safety, thus lowering the rate of injury.

Our major comparison for physical education statistics comes from the National Safety Council of Chicago, Illinois. In a 1977-78 survey of over 22,000 school districts, they reported a rate of 2.4 accidents per 100,000 days. The Safety Council advised us to translate this into a million student hours by multiplying by 10 (100,000 to 1,000,000). This computation results in a rate of 9.6 injuries per million student hours of exposure to regular physical education classes. These injuries all were defined as those resulting in a day's loss or more from school.

Linda Higgins, in a "Physical and Sports Medicine" article, gives the following statistics for further comparison. According to her research, Outward Bound has a rate of .9 injuries per 1000 days of program exposure. Even if we assume 24 hours per day of exposure for Outward Bound courses, this translates into a rate of 37.5 per million hours.

In the same article, Linda Higgins quotes a statistic for automobile driving, which (when translated into million hours) results in a rate of 60 injuries per million hours.

To review the comparable statistics:

Program Exposure hours	Accident Rate per Million
Project Adventure (15 yr)	3.67
Project Adventure (10 yr)	5.13
Physical Education Classes	9.6
(National Safety Council study)	
Outward Bound (L. H. article)	37.5
Automobile Driving (L.H. article)	60

Analysis of Injuries by Activity

The attached Exhibit B shows the frequency of injury by activity. The trend of our first study, which showed that most injuries occurred on the low elements, has clearly continued. Over 85% of the reports related to injuries on low, unbelayed activities (initiatives, games, warm-ups). The majority of these injuries were muscle pulls or sprains as a result of sudden movement or lack of proper spotting. The more severe accidents on the low elements involving broken bones or separations were a result primarily of those low elements that necessitated active spotting, and where often either passive spotting or participant inattention to instruction was involved.

Injuries on the high elements were primarily a result of climbing movements that caused strain to a limb or joint, resulting in a sprain or muscle pull. Typical of this type of injury would be this report, "Participant strained shoulder attempting to regain position on the Heebie-Jeebie." There were three incidents reported in which participants on belay were allowed to descend too quickly by their belayer, and the proper attentive techniques by the belayer could have prevented the problem. For both the high and low elements, there were no injuries reported, to our knowledge, that resulted in significant permanent disability.

The top three activities in frequency of injury are the Electric Fence, the Wall, and the Zip Line. In each case we have initiated changes which should reduce the frequency of these events.

Electric Fence

This is a low initiative problem in which participants have to help each other over a rope 4-5' high, using a foot beam. It is a popular and effective group building initiative, but one that demands attentive spotting. Since our last study, which also had the Electric Fence as the activity with the greatest frequency of injury, some programs have dropped this activity. Reluctantly, Project Adventure Inc., has decided to discontinue this activity in our instructor training workshops. Our reasoning is that there are many other activities with good group-building potential and that are much less difficult to spot. We suggest programs which use this activity re-evaluate their policy on this activity. If they continue to use it, we suggest careful attention to spotting.

Wall

This is also a popular activity of long standing and also requires active, attentive spotting. However, half of the accidents on the Wall were a result of the decent from the rear of the Wall and most of these resulted from participants jumping the last 2-4' unassisted and either spraining or breaking an ankle in the process. Obviously, proper spotting of the rear descent is important, and staff should remember to instruct participants to climb down slowly. Many programs now utilize a platform with either a protective rope, or a net on the rear of the Wall. This platform, in conjunction with a ladder, is an option many programs are now using with success. With the use of a rear platform, or with more careful spotting of the rear descent, the incidence of injury on the Wall should decrease. Instructors should also remember that a rear platform still requires spotting.

Zip Line

More than half the injuries on the Zip Line were a result of the use of a participant-held braking system that Project Adventure has not used in training workshops for a number of years — the "trust brake." The injuries (muscle pulls, bruises and sprains) were primarily to participant "brakers", who in most cases were not sufficiently prepared for their task. Project Adventure now recommends that Zip Lines utilize either a gravity braking system or a mechanical system such as Braking Tires or the PA Brake Block system. These systems are both time-tested by now and have involved no reported braking injuries. Their use should significantly reduce the incidence of injury on this element.

Equipment Related Issues

There were three significant injuries reported as a result of equipment problems. One involved an injury to an instructor as a result of a faulty installation of a Zip Wire. The installation was by a volunteer group and had not been inspected by professionals in the field. The second injury was a result of a participant falling 4-5' when a rung on the back of a Wall pulled

out. Again, this ladder was constructed by volunteer help and was not inspected by an outside group.

One of the Zip Line injuries was a lacerated knee as a result of a Zip Line that was constructed too low, and was not inspected by professionals in the field.

Conclusions

The surveys returned covered over one million participants over the past fifteen years. The data covering these participants allows us to put into proper perspective the safety record of PA programs. This study only serves to reinforce the conclusion of our Ten Year Study: a Project Adventure program is as safe as, or safer than, a more traditional physical education program.

The safety record of the program is a result of two major factors. First, the systems of belaying and spotting procedures are now time tested, reliable systems. They are reliable because of both the sound original design of the systems and their refinement over time. Secondly, to date there has been a consistent and growing professionalism of the teachers, counselors, trainers, and administrators who run programs.

To continue our good record in the future as the number of programs continues to grow, the following are recommended:

- The single most important criteria in program safety is the quality of training of staff. Proper training by professionals of all staff in both program skills and program design is necessary.
- Regular inspection of equipment, both by internal staff and by outside professionals, is necessary for proper maintenance reviews.
- Regular staff retraining and outside review of program skills is a key to maintaining staff quality and program safety. The recently developed on-site Project Adventure Accreditation Program is proposed to meet this increasing need. The application of the standards of this Accreditation Program should be a factor in continuing our good record for what is an increasingly significant field of education.

Exhibit A

Project Adventure, Inc.

February 1986

Dear Friends,

We need your help!

Project Adventure completed a Ten Year Safety Study of Project Adventure type programs using a Ropes Course in 1981. The results of that survey, as many of you know, showed the activities to have an incidence of injury less than that of standard physical education classes.

Now, as we enter the fifteenth year of Project Adventure programming, it is important to do a follow-up study. The tight liability insurance market of 1985 has brought increased pressure on Adventure programming, and some programs have had insurance problems. One of the answers to this is to build a reliable data base and to be able to argue from the facts.

In 1980, we mailed to 246 sites and had a return of 116 (47%). This survey is being mailed to over 700 sites. These are all programs that we have contact with in some way, and that have an adventure program using a Ropes Course.

Please take time to fill out the survey and return to us in the enclosed envelope. The strength of the adventure programming field rests, in large part, on our being able to assure the public of safe, quality programming.

If you answered our 1980 survey, please answer this one with the facts from day one of your program. We will compare the results ourselves with the previous survey.

We are sending this survey with the help of the Rhulen Agency of Monticello, New York. We are currently working with Rhulen to help insurance companies understand our activities. We have hopes of establishing a group program for liability insurance for those programs in need. The philosophy of group problem solving and collaboration that we all subscribe to is being put to a crucial test during this period. Thanks for your assistance in filling out the form.

If you wish to receive a copy of the completed survey, please be sure to enclose a stamped, self-addressed envelope with your completed survey.

Thank you, again, for your cooperation.

Sincerely,

Richard G. Prouty

Project Director

Project Adventure, Inc.

Exhibit A cont.

Questionnaire

1. Site (Name of Institution)______________________________
   Address______________________________
   Phone#__________________
   (zip code)__________(area code)____________
   Main Contact Person ______________________________
2. Do you have an Adventure Program using a Ropes Course? Yes__ no __
3. Low or High Elements, or both?______________________________
4. How many of each? ______________________________
5. Which of the following best describes the type of Adventure Program at your site? (check one)
   A.______PA type physical education program
   B.______PA type Adventure Based Counseling program
   C.______Camp program
   D.______Other — Please describe briefly______________________
   ____
6. Approximately how many persons will participate in your adventure program annually? ______________________________
7. What is the approximate length of each participants program? (example — one six month period, four hours a week)______________________
   ____
8. If varied length, please estimate average participant days or hours:______
9. How many different instructors are leading activities at your site?______
10. What is the average ratio of staff to students?__________________
11. Where was Ropes Course training acquired by your staff?____________
    ____
12. Was the equipment installed by an outside agency or person?___yes___no
    ______________partially?__________by whom?____________
    What was the date(s) of installation of equipment?______________
13. Do you have an outside agency inspect your equipment?_____yes_____no
    How often?________Who has done this for your site?______________

14. The following information on your liability coverage will be helpful. Your business manager, superintendent, or director will usually have this information.

    A. Name of insurance agent________________________________

    B. Name of insurance company_____________________________

    C. Amount of general liability coverage (i.e., $500,000, $1,000,000, etc.)

    ____

    D. Cost of insurance (if specific to your adventure program).

    ____

15. Please list on a separate page all significant injuries occurring as a result of participation in your Project Adventure program and/or Ropes Course activities. A significant injury is one that resulted in the loss of time from school or work by the participant. Please include a brief description of the time of the accident.

16. Have there been any legal claims as a result of an injury in your program? If so, please describe the status of legal action.

    ____

    ____

17. Please use the space below for additional comments, questions, or concerns you may have relating to safety and/or insurance issues affecting your adventure program.

Exhibit B

Activities (w/more than 1 injury reported)	# of injuries
Electric Fence	19
Wall	18
Zip Wire	11
Beam	6
Trust Fall	6
Rope Swing/Seagull Landing	5
Tension Traverse	5
Mohawk Walk	4
Hickory Jump	4
Climbing Wall	4
Flea Leap	3
May Pole	3
Swinging Log	3
Track Walk	3
Two Line Bridge	2
Pamper Pole	2
Cage Ball	2
	100

No other activity had more than one injury reported. The list of single injury activities includes most of the initiatives, games, and warm-ups, as well as some other "adventure activities," such as hiking, cross country skiing, sledding, etc.

57 activities, for a total of	57
	157

(Total injuries reported for
1.2 million participants over 15 years)

Project Adventure, Inc.

Appendix D

Rope Use Log

It is important that programs keep track of rope use. Pertinent information should be readily available to program instructors and staff, and upon a course inspection. This information includes age of the rope, use of the rope in number of days, and the average number of participants using the rope during a given time period. Keeping track of all groups going through a course during a month's time, times the average number of participants per group, will give a reasonably accurate count for a rope being used full time during that period. An easy way to identify a rope is by wrapping a piece of tape around an end and writing on it the date placed into service and some identifying mark; e.g., letter or number. There are many ways to keep track of this information, but the most important thing to remember is that the information is available and is accurate. The rope use log is one way to keep a record of use. Blank spaces have been included for individuals to add information they also feel is important, such where the rope was purchased, type of rope, date into service, etc.

Rope #	**Age (date into service)**

Days in Use	**Ave. # of participants per day**

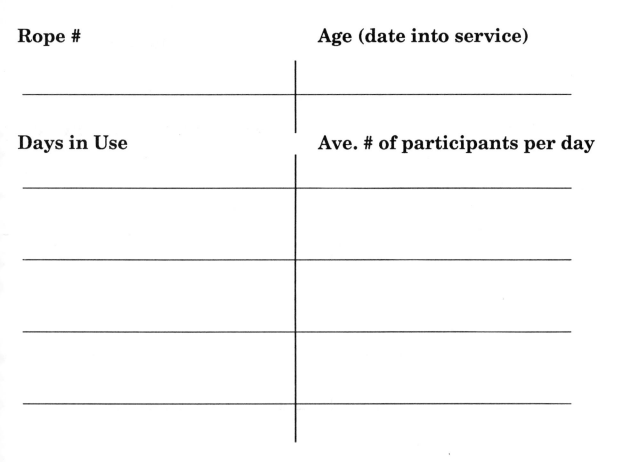

Rope #	**Age (date into service)**
Days in Use	**Ave. # of participants per day**

Rope #	**Age (date into service)**
Days in Use	**Ave. # of participants per day**

Your Comments Please

Because this manual is the result of a two year project that involved the input of many different people, we would like to solicit feedback from you folks in the field. No one knows better what is needed to ensure that programs continue to be run safely and effectively, and how this manual helps to that end. So — let us hear from you. Use this page to offer suggestions, criticisms, items for inclusion in the next edition, helpful insights, etc. We'll listen.

Mail to:

Project Adventure, Inc.

P.O. Box 100

Hamilton, Ma 01936

Project Adventure, Inc.

P.O. Box 100 • Hamilton, MA 01936 508–468–7981
P.O. Box 2447 • Covington, GA 30209 404–784–9310
A NON-PROFIT CORPORATION

Since 1971 Project Adventure has been creating learning programs that challenge people to go beyond their perceived boundaries, to work with others to solve problems and to experience success.

The Project Adventure concept is characterized by an atmosphere that is fun, supportive and challenging. Non-competitive games, group problem-solving initiatives and ropes course events are the principal activities we use to help individuals reach their goals: to improve self-esteem, to develop strategies that enhance decision-making, and to learn to respect differences within a group. Our mission is to help people adapt Project Adventure programs where they live and work — to "Bring the Adventure Home."

KENDALL/HUNT PUBLISHING COMPANY
Dubuque, Iowa

ISBN 0-8403-6207-2